PHYSICS
AROUND THE
CLOCK

Praise for Physics Around the Clock

'*Physics Around the Clock* explains, in an easy and engaging way, how from morning till night, we're surrounded by fascinating physics that hides in plain sight.'

James Kakalios, physics professor at the
University of Minnesota and the author of
The Physics of Everyday Things

'Physics is all around us – even as we go about our seemingly mundane daily lives, as Michael Banks ably demonstrates in *Physics Around the Clock*. Whether it's your morning coffee, daily commute, walking the dog, cooking dinner, playing Monopoly or Texas Hold 'Em, or debating whether it's better for gunslingers to draw first while watching classic spaghetti Westerns, a physicist somewhere has studied it. And Banks is here to explain it all to you in a truly compelling read.'

Jennifer Ouellette, author of
The Calculus Diaries

'This is one of those brilliant books where there is an amazing fact on every page showing just how much science underpins our everyday world.'

Mark Miodownik, author of *It's a Gas:
The Magnificent and Elusive Elements
that Expand our World*

PHYSICS AROUND THE CLOCK

ADVENTURES IN THE SCIENCE OF EVERYDAY LIVING

MICHAEL BANKS

For Claire, Henry and Elliott

First published 2025

The History Press
97 St George's Place, Cheltenham,
Gloucestershire, GL50 3QB
www.thehistorypress.co.uk

British Library Cataloguing in Publication Data.
A catalogue record for this book is available from the British Library.

ISBN 978 1 80399 582 3

Typesetting and origination by The History Press
Printed and bound in Great Britain by TJ Books, Padstow, Cornwall

MIX
Paper | Supporting
responsible forestry
FSC
www.fsc.org   FSC® C013056

The History Press proudly supports

Trees for Life

www.treesforlife.org.uk

EU Authorised Representative: Easy Access System Europe
Mustamäe tee 50, 10621 Tallinn, Estonia
gpst.request@easproject.com

CONTENTS

INTRODUCTION

IF I ASKED YOU to outline a typical twenty-four-hour day, what would you say? Perhaps you would describe sleeping for roughly eight hours each night, with seven hours spent working or studying and then whatever is left squandered on commuting, chores or perhaps taken up with looking after children. If you are lucky then you may mention having a few extra hours for yourself before bed to watch TV, play a board game or read a book, such as this one. While you might think that your day is unique, it turns out that a large swathe of the world's population spends a similar amount of time doing the same sort of tasks. At least, that is according to researchers at McGill University in Canada, who in 2023 came up with a typical 'global human day'.[1]

To do so, they gathered data taken between 2000 and 2019 by national statistics agencies and international organisations, representing fifty-eight countries, or about 60 per cent of the world's population. Once they crunched through all the numbers, they discovered, somewhat unsurprisingly, that sleep, or bedrest, makes up the largest single chunk of 'activity', occupying – and those with young children look away now – about nine hours each day.* The remainder, or some fifteen hours, was then split between three categories: the largest, with a similar amount of time as sleeping, was spent on 'human outcomes', such as making yourself look presentable, caring for children, reading, watching TV, playing sport, going for walks, and just generally chilling out; another three hours were consumed by activities on 'external outcomes' like food preparation, as well as cleaning and tidying the house; the remaining two or so hours were taken up with socialising or commuting.†

Many of the activities that we carry out day in, day out, we do so at the same time and for the same duration (especially sleeping) each day.‡ You probably have breakfast within the same ten-minute window every morning, have a shower just before 8 a.m., and leave for work, take the kids to school, or the dog out for a walk thirty minutes later. We are creatures of habit, after all. While some of these deeds may be considered rather ordinary and even downright boring,

---

* Admittedly this number is high because it also includes children, who generally sleep longer than the eight hours a day that an adult may sleep.

† You might wonder why 'work' or 'study' is missing, but the authors already included work activities into the other time categories (such as food production). On top of that, as it is an average global human day, the number of hours spent 'working', averaged across all humans, is only 2.6 hours per day.

‡ While the 'global human day' is remarkably consistent across people of different countries and backgrounds, there are a few cases where it isn't. Those from low-income countries, for example, spend more time on farming or food preparation than those in higher-income nations.

taking the time to stop for a moment and examine what is happening underneath your very nose (or eyes) can reveal a whole host of surprising and extraordinary phenomena. Taking the dog out for a walk in the rain, for example, might be a pain, especially when the pooch comes back inside the house and carries out a 'wet dog shake'. Yet the ability to make water droplets fly off in every direction can be a matter of life and death for your mutt, helping to stave off hypothermia. Modern high-speed photography has revealed that the movement is so effective it can eject some 70 per cent of water from a dog's fur in just a few seconds, going some way to explain why you, and the furniture, get so wet in the process (for more, see Chapter 4).

That is what this book is about: delving a little deeper into the seemingly ordinary aspects of everyday life to reveal the fascinating science and physics that lay beneath. Doing so reveals how the beauty and versatility of physical laws can manifest themselves everywhere, whether it is the way that Cheerios* mysteriously clump together when floating in a bowl of milk at breakfast to how opening a bottle of champagne (presumably not in the morning) results in a supersonic plume of gas – the same physics that occurs in the exhaust of jet fighters. Indeed, searching for interesting or new phenomena does not always require building billion-pound experiments, such as underground particle colliders or launching huge astronomical observatories into the depths of deep space. While no one can deny that particle smashers, such as the Large Hadron Collider at the CERN particle-physics laboratory near Geneva or NASA's

---

* Other ringed-shaped breakfast cereals are available.

James Webb Space Telescope, have done, and continue to do, amazing science, such endeavours often attract the most media attention, resulting in what most people take physics to be. Instead, in the pages ahead we will examine the more familiar world around us, whether in the simple sound of dripping water into a water-filled container to how a dandelion pappus drifts effortlessly for metres in the wind to turn a lawn into a summer meadow.

We begin by exploring the morning routine and the intriguing physics that can be seen as we have breakfast. As the typical 'global human day' shows, commuting is a significant part. We will also delve into the interesting dynamics involved with getting around town, as well as investigate the science involved in other common daily pursuits, such as enjoying the garden. Finally, we end with the physics behind popular evening pursuits, such as cooking dinner, playing your favourite board game or the network theory behind your favourite movie. Each chapter is self-contained, and doesn't necessarily follow the other, so if you have no interest in ball sports, or the spread of pathogens makes you feel a bit icky, then simply move on.

Along the way we will meet the scientists who find fascination and intrigue in examining the how and why in everyday phenomena. Their discoveries highlight how following your curiosity can reveal countless marvels that exist in the mundane. Even during the rush of our daily lives.

1

WAKE UP AND SMELL THE PHYSICS

WHAT BETTER WAY TO start the day than with a nice, warm cup of coffee – especially if it is brought to you as you lay in bed. If you are lucky enough to receive that level of service, then you will already know what is coming, thanks to the aroma of the cup of java as it nears. There are thought to be more than a thousand chemical compounds that are extracted from ground coffee, which give the drink its unique flavour and smell. Yet taste isn't everything on offer: it's the $C_8H_{10}N_4O_2$ that really counts.* Coffee is the beverage brewed from the roasted and ground cherries, or beans, of the coffee plant. Although there are more than a hundred coffee plant species that have been catalogued to

---

* This being the chemical formula for caffeine. Of course, some people prefer or can only drink decaffeinated coffee. This can be achieved in several ways, but one method is to steam the green beans before soaking them in a solvent, such as methylene chloride, which removes about 97 per cent of the caffeine. The beans are then dried and roasted.

date, only two varieties are used commercially: *C. arabica*, or Arabica coffee, and *C. canephora*, which is commonly known as Robusta. Arabica, which accounts for roughly two thirds of global coffee production,[*] is thought to have a sweeter and smoother taste than Robusta, but comes with the downside that it can only be grown in a few places in the world where climate conditions are favourable.[†] Robusta, as the name suggests, is more resistant to disease and adverse weather conditions than its Arabica cousin.

A cup of coffee is the result of growing, harvesting, roasting and grinding coffee beans before passing water through the grounds to extract all that goodness. The first flavour that emerges when roasting beans is acidity but thanks to a chemical process called the Maillard reaction,[‡] chocolatey, nutty or caramel flavours soon emerge. What you will likely be in control of is how to brew the coffee itself (if you happen to be a coffee connoisseur then you may also grind

---

[*] Arabica is thought to have been created naturally from two parent species – *Coffea canephora* and *Coffea eugeniodes* – around 600,000 years ago in the forests of Ethiopia. This is some 300,000 years before modern humans, refuting the claim that humans bred the species. The amalgamation of the two parent plants is thought to have resulted in Arabica's flavour and its large and complex genome.

[†] Some estimates predict that, by 2050, suitable areas for growing Arabica will fall by about 80 per cent due to climate change, see Imbach, P., Fung, E., Hannah, L., et al., 'Coupling of Pollination Services and Coffee Suitability Under Climate Change', *Proceedings of the National Academy of Sciences*, vol. 114, no. 39 (2017): 10438–10442.

[‡] Named after the French chemist Louis-Camille Maillard, who first described the process in 1912. The reaction between sugars, heat and amino acids is also responsible for giving certain foods their distinct flavour – think toasted marshmallows or seared steaks.

– or even roast – your own coffee beans. Fancy!). Coffee can be made several ways , with each way having a slightly different technique and differing results. Some enjoy the simplicity and ease of instant coffee for a quick caffeine fix.*
Instant coffee granules can be manufactured in a few ways, but one method involves brewing coffee beans and then evaporating the resulting liquid at a high temperature and low pressure. The remaining condensed extract is freeze-dried at -40°C and broken up into small crystals. These are the ones that you scoop into your cup before adding hot water, and perhaps a splash of milk and some sugar. While a cup of instant coffee may offer a quick and easy caffeine fix, it only contains about 70mg† of caffeine per cup compared to some 80–100mg for an espresso shot.

Over the past twenty years consumers have not only become more conscious about where their coffee comes from but also more precious (perhaps snobbish?) about how it is made. For decades, I was easily content with the ease of instant coffee, perhaps branching out to filter or a trip to the local coffee shop at the weekend for an espresso. But I never looked back when I bought a 'bean-to-cup' espresso coffee machine just before the COVID-19 pandemic, which resulted in coffee shops temporarily closing during the lock-downs. Since then, I have enjoyed the ease each morning (and afternoon) of pressing a button, hearing the beans being ground, followed by a nice caffeine hit. What I never quite appreciated, however, was the physics that goes into making

---

* The amount of caffeine in your bloodstream peaks about fifteen to forty-five minutes following consumption.

† A milligram being 0.001g.

the perfect cup of coffee, requiring the precise combination of water temperature, flow rate and sometimes pressure to get the balance of flavours just right. So, tossing aside instant coffee, how can physics help to make that perfect cup?

If I said that volcanic eruptions and grinding coffee beans have something in common, the first thing that might come to mind is being too eager to drink from your takeaway coffee cup and scolding your mouth and tongue. In fact, both processes involve the production of a lot of static electricity. Static occurs due to an imbalance between negative and positive charges on an object as it is rubbed together with another object. These charges build up on the surface of an object until they find a way to be released or discharged – the classic experiment being to rub a balloon on someone's hair. The hair will become positively charged by losing electrons, while the balloon will be negatively charged, gaining electrons. Bringing the two together results in the hair rising towards the balloon as they attract.

Grinding coffee involves fracturing the beans via rotating metal plates in a burr grinder.* This grinds the beans to a fine ground but, as it does, it produces static, with individual coffee grains acquiring an electrical charge. The static produced can be high enough to cause the small coffee grains to clump together as well as stick to the grinder. To get around

---

* Burr grinders involve two revolving abrasive surfaces that grind the beans, while blade grinders use a propeller-like blade akin to a blender. Burr grinders produce a more consistent grind than blade grinders.

this, some baristas add a drop of water to the beans before grinding to reduce the static friction – a method known in the industry as the Ross droplet technique. Not much was known about how this process impacts the resulting brew until 2023, when Christopher Hendon at the University of Oregon in the US teamed up with volcanologists – yep, people who study volcanoes – to measure the static electricity produced when grinding roasted coffee beans.[1]

Hendon, who goes by the moniker 'Dr Coffee', has spent his career studying coffee and first became interested in the tipple when doing a PhD at the University of Bath in the UK. He was surprised to find that while there are certain rules when it comes to coffee-making (more on that later), not much scientific research had been done on what goes into making the perfect cup. 'What sets coffee apart from many other drinks is that you have to do something before the point of consumption,' Hendon told me. 'You can open a bottle of beer or wine and drink it how it is intended to be, but for coffee you must make it. From that perspective, it's inherently an experimental beverage because it entices the consumer to play around with the variables until they arrive at something they think tastes good.'

Back in the US, Hendon and colleagues began running public coffee-making demonstrations at Oregon University, when a couple of volcanologists came by to sample what was on offer. They became intrigued about the static that was produced when the coffee beans were being ground. After discussing this with Hendon, they became, for want of a better word, attracted to the idea of investigating the charge that can build up during bean grinding. The experiments the team carried out involved modifying an instrument known as a 'Faraday cup', which is used to measure the electric charge of volcanic ash. This is a material that can be so electrically charged as the particles collide and fragment that when it gets spewed into the atmosphere it can even cause volcanic lightning storms. The result of their efforts was a small metal vessel – about the size of an espresso cup – that the team placed under a burr grinder and used it to measure the static charge of individual grains as they were ground. They found, as predicted, that the charge on the grounds was larger the more finely the coffee was ground. What wasn't so expected was that lightly roasted beans had a positive charge while darker roasts had a negative charge. Unroasted beans contain about 50 per cent water, but the roasting process removes liquid so that dark roasts only contain about 10 per cent water. The higher the internal moisture content of the beans, the lower the charge on the grains, suggesting that water reduces the static by acting as a lubricant during grinding.

When Hendon and colleagues then compared espresso made with identical coffee beans that were either ground with or without a drop of water, they discovered that grinding with water resulted in a more consistent and stronger brew. They attribute this to water reducing the friction during grinding and thus stopping the grounds from clumping. This allows the water to move more consistently, and more slowly, through the uniformly compacted grounds during brewing to help extract more flavour. Hendon and colleagues discovered that adding a splash of water resulted in 10 to 15 per cent more 'yield' – the fraction of the ground coffee that dissolves and ends up in the final drink – compared to not adding a drop of water.

So, if you like to grind your own beans at home, perhaps think about adding just a drop or two of water before grinding to stop the grains from clumping, which will hope-fully result in a tastier coffee.

Brewing coffee from ground coffee beans is all about hot water moving through a bed of coffee grains so that it can absorb the flavours and oils of the coffee as it passes through. The basic physics of what happens when water percolates through a bed of granular* material such as ground coffee was first worked out in the nineteenth century by the French hydrologist Henry Darcy. In the 1840s, Darcy was con-tracted to provide a clean supply of water to the French town of Dijon, which a few years later would become known for its mustard. Between 1855 and 1856, Darcy undertook a series of experiments in which he tested how water was filtered as it flowed through a cylinder of sand. He had to calculate how many cylinders would be required to filter the necessary amount of water, and from the experiments he empirically† obtained what is now known as Darcy's law, which describes the speed of a liquid as it moves through a bed of granular material. Darcy's law has had countless applications in everyday life, such as describing how water, oil or gas flow from petroleum reservoirs – and, of course, in coffee extraction.

Darcy's law states that the flow rate of a liquid depends on several factors, one of which is the viscosity of the liquid.

---

* Granular as in resembling or consisting of small grains or particles.

† This means that the law was used to describe the results of experiments rather than being based on theory or logic.

This is the resistance that a fluid has to a change in shape or movement. A fluid with a large viscosity, such as honey, resists motion because its molecular make-up produces a lot of internal friction. A fluid with low viscosity, like water, flows easily because its molecular make-up results in very little friction when it is in motion. If you have a cup of water and a cup of honey and tip both, the water, with its lower viscosity, will flow faster. Other factors that affect the liquid's speed through a granular bed include the pressure difference between where the liquid enters and where it exits. Yet when it comes to applying Darcy's law to brewing coffee, the key parameter is the permeability of the granular bed of coffee beans. A material's permeability is a measure of how easily it lets liquid through. When it comes to coffee making, it can be changed by compressing the coffee grains into a firm bed, which is why you will often see baristas 'tamp' their coffee into a compressed puck before putting it in the machine. This helps to pack the grains together so that water can't find an easy escape route through the puck, which would result in a weak brew. Another way to alter the permeability is to change the size and shape of the grains. Finer grinds maximise the surface area of coffee that is in contact with the water, which means it can take longer to pass through. But if the bed is compressed too much, then it can take too long and this can result in extracting higher levels of organic acids, leaving a bitter taste. 'Making coffee is about extracting the molecules that taste good as fast as possible,' adds Hendon.

So how does this work in practice? The simplest coffee-making contraption from a physics perspective, and one that is found in countless domestic kitchens, is the French press. Here, hot water is poured on to coarse or large irregular-shaped coffee grains that are held in a cylindrical container. In this 'immersion' scenario, the water is always in contact with the grains, leeching the coffee compounds into the liquid. Over time, the grounds rise to the top of the liquid to reside under the base of the plunger, which is then used to push the grounds back through the water. If you have ever unknowingly scooped coffee grounds into a French press without any feel for how much is the right amount, then the recommended method is 54g of coffee grounds for 1 litre of boiling water. If the above recipe is always used then the only parameter that can be changed is the time the water is in contact with the grains, which is usually suggested to be five minutes before plunging. And if you have ever wondered how much force is needed to push the plunger through the water, in 2021 physicists carried out a combination of kitchen- and laboratory-based experiments to determine that 32 Newtons[*] is the magic number. This force took the plunger in a 1 litre French press from top to bottom in about fifty seconds.[2] Now you know.

Things get a little more complicated for filter coffee, which remains a popular way to brew despite the rise of espresso machines. This method involves putting the coffee grounds into a conical paper or metal filter and then pouring near boiling water from the top. The water slowly percolates through the coffee, thanks to gravity, over several minutes and as it does so the water washes over the coffee grains, extracting some of the flavour before passing through the filter and dripping into a glass container placed on a heated plate. The key parameter

---

[*]   Newton is the unit of force. About 32 Newtons is the same force a
      2-year-old child can exert when pushing with their thumb – although
      I don't recommend that you try this using a small child.

here is the time it takes the water to wash over the coarse coffee grains. A finer grind makes the water seep through more slowly, so it can increase the extraction, but it also results in more unwanted bitter compounds dissolving into the water. Filter coffee is Hendon's preferred method in the morning, which he makes thanks to a Moccamaster automatic that helps to automate the process of adding water. 'It saves me the five minutes it would take to pour water through the beans,' he adds, 'and it makes good coffee, well, good enough for me.'

Our physics-based coffee-brewing venture gets a bit more interesting when some vapour pressure is thrown into the mix. This happens in the Italian 'moka', which was invented in 1933 by the Italian engineer Alfonso Bialetti and was widely commercialised in 1946 by his son Renato via the trademark Moka Express. In a moka, which is still prevalent in Italy today, a bottom container is filled with water and on top sits a basket that holds finely ground coffee beans (the grounds should be packed loosely and not compressed). The bottom compartment plus basket is then screwed on to an empty upper chamber, which is often in the shape of a teapot (see figure opposite). The moka pot is then put on the stove and is heated from the bottom compartment. The heat gradually turns the water into steam, the pressure rises and in turn the steam 'pushes' the water up the middle funnel and through the coffee grains in the basket and into the top teapot compartment above.*

---

* In 2012, physicists used a beam of subatomic particles called neutrons, which were produced at the Paul Scherrer Institute in Switzerland, to take real-time images of what was happening inside the moka pot as it was heated. For a video, see youtu.be/VESMU7JfVHU.

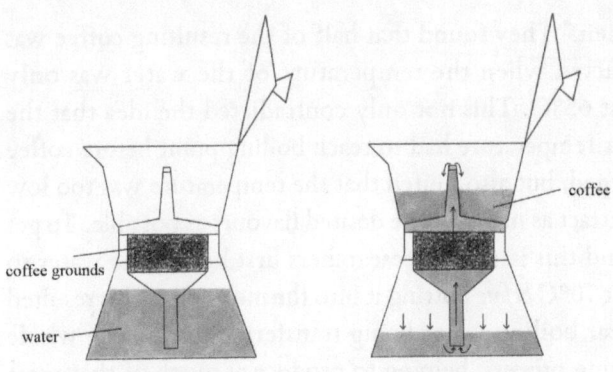

In a moka pot, the pressure from the steam drives the hot water up through a bed of coffee grains and into the top compartment.

By measuring the flow rate through a bed of ground coffee, in 2007 scientists applied Darcy's law to the moka, finding that the coffee bed's permeability is akin to sandy silt or clean sand (showing that Darcy's experiments with sand columns does have something in common with brewing coffee).[3] Yet two years later another moka pot study found that things are slightly more complicated, and the permeability of the coffee bed was not constant over time but reduced as the pressure of the water and its flow rate changed during the brewing process.[4] Despite being a seemingly simple device – what some might call a 'poor person's espresso' – the physics of what is going on inside a moka pot is far from straightforward. So, how can you improve the output? In 2008 (the physics of the moka pot was obviously a hot topic in the late 2000s), physicists carried out controlled experiments in the lab. They first sealed a moka pot with water at room temperature and heated it by about 10–20°C per minute – what typically happens in a domestic kitchen.* They found that half of the resulting coffee was extracted when the temperature of the water was only about 65°C. This not only contradicted the idea that the water temperature had to reach boiling point before coffee emerged, but also hinted that the temperature was too low to extract as many of the desired flavours as possible. To get around this issue, the researchers first heated the water to about 70°C *before* putting it into the moka pot. This resulted in near boiling water being transferred during the whole brewing process, helping to produce as much of that great flavour as possible. If you rely on a moka pot for your morning coffee, next time perhaps try starting off with hot water and see if it tastes any better.[5]

Perhaps the ultimate way to prepare coffee – especially for busy coffee shops – is the espresso† machine. In simple terms, water is boiled to produce steam, and this is collected in a 'group head' at the top of the machine that pressurises the steam and uses it to drive water through the bed of grounds that are held in a portafilter, which then drips into a cup held below it. Espresso normally comes in a fine grind and the Specialty Coffee Association – a non-profit, membership-based organisation that represents thousands of coffee professionals around the world – defines an espresso as a 25–35ml beverage prepared from

---

* Another tip is to make sure the safety valve on the bottom of the device is functioning. If the coffee bed is strongly compressed, meaning the water cannot escape and the steam safety value doesn't open, the pressure inside can turn the moka pot into a bomb.

† Espresso meaning 'quick' in this context.

about 7–9g of ground coffee.* The association states that water should be heated to 92–95°C before being forced through the granular bed under 9–10 bar of static water pressure for between twenty to thirty seconds.[6] If the water is hotter, it risks burning the coffee grounds, resulting in a bitter brew, whereas if the temperature is much less than 90°C then – as we have seen what can happen with the moka pot – the flavours will not be extracted efficiently, resulting in weaker coffee.

In a filter or other 'pour over' coffee technique, about 1.5 per cent of the coffee material is dissolved in the beverage, but for espresso it is about 10 per cent – so almost ten times more concentrated. While adding water to an espresso shot results in the same concentration of coffee mass as a filter or coffee made with a French press, it will taste different, being perhaps more intense. This is because many molecules in coffee are volatile – easily lost to the atmosphere – but adding pressure, as in the espresso technique, helps to lock those molecules inside the resulting drink. The addition of pressure driving the water through the grounds is also thought to help extract more of the rich oils into the resulting coffee. Indeed, if you do it right then you will see the result – the 'crema', a golden-brown layer of tiny bubbles of carbon dioxide that are coated in proteins and oils of the coffee.

The espresso is widely seen as the most complex way to make coffee – Hendon calls it 'a high-performance version of filter coffee'. But with that complexity brings a certain

---

* Coffee shops in many countries use a higher coffee mass of 15–22g, resulting in larger-volume beverages of 300ml, much to the dismay of Italians.

susceptibility to inconsistency. While espresso machines control the temperature and pressure of the water, the rest is left to the barista.* Even for skilled baristas operating top-of-the-range coffee machines, creating cups of espresso that taste consistent from one to the next can be tricky. The coffee industry in the US is worth $343 billion, or 1.5 per cent of US gross domestic product, which means that any improvements to the process can have a big impact – as we have seen in the Ross droplet technique by simply adding a drop of water before grinding.

As espresso machines, grinders and beans are different, that means there is, unfortunately, no magic formula to tell you the right parameters to use at home. Coffee is a tricky drink to prepare because there are so many factors that can be changed from the temperature and amount of water to the coarseness of the beans and how compact they are. 'My main tip when preparing coffee is that every single decision you make, either consciously or unconsciously, will impact extraction,' notes Hendon. 'The first thing you can do is try and measure everything you have done, whether it's the brew time or temperature of the water to better understand why you're getting that particular output.'

This dependence on experimenting means that coffee is perhaps the perfect physics-based drink, and goes some way to explain why it is the beverage of choice for physicists. Yet a little knowledge of what is going on when making

---

* A study in 2021 found that coffee is perceived to taste sweeter if it features latte art; see Hus, L. and Chen, Y-L., 'Does Coffee Taste Better with Latte Art? A Neuroscientific Perspective', *British Food Journal*, vol. 123 (2021): 1931–1946.

that morning tipple can also help you to deliver the best possible coffee.

You might think that once you have finished your cup then that is where the physics ends. Well, there is one last thing. It is always hard to keep any liquid from trickling down the side of a mug as you take your mouth away after a sip. If you leave your cup in one place for a while you may see a ringed stain left behind. After all, it's why we have coasters. Likewise, if you leave a couple of drops in your finished coffee and let them dry, you may see that the droplets looks like a ring – lighter in the middle with a dark ring around the outside. This effect can be seen for other types of drinks, such as wine, and it is likely to have been observed hundreds of years ago too, but it is only in the past couple of decades that it has been made clear what was going on. It is now known – rather unimaginatively – as the coffee-ring effect.[7]

If you take a droplet that contains tens of little dark particles inside and watch it dry under a microscope, you will see that the particles all seem to move to the drop's edge, where they collect, resulting in a dark ring as the drop dries. The question is, why are the particles attracted to the outer edge rather than just being spread evenly over the drop's surface? The explanation is that as a drop evaporates, the edge of the drop stays pinned. While the drop stays the same diameter, the height of the droplet decreases as it evaporates, so it ends up a bit like a pancake, or a spherical cap. The evaporation does not happen evenly but is quicker at the edge of the drop. As this happens, liquid at the edge is replenished by the interior, and this creates a current towards the edge. The particles inside the drop then flow outwards to the edge, where they pile up, one at a time, into a tightly jammed packing, which produces a dark ring as the drop finally dries.

You might think that the coffee-ring effect is just an annoying aspect that must be cleaned, but knowledge of it can also have lifesaving applications. Scientists are using the effect to design simple and cheap ways to diagnose diseases, such as malaria, which kills more than a million people a year.[8] In such a test, a patient's blood is mixed with a liquid containing gold nanoparticles and then left to dry. If malaria is present in the blood, then a unique protein* that is secreted by the malaria parasite will be present. This protein then binds with the nanoparticles to form clusters and when the drop dries these clusters move to the edge, leaving a visible ring. If no malaria is present, then the clusters don't form and there is no ring.

A simple test for malaria all thanks to the same physics that happens with your finished cup of coffee.

---

* Called histidine-rich protein II, or HRPII.

2

BREAKFAST
WITH EINSTEIN

WHILE THE UK HAS a growing love for coffee, it might not be the first drink you consume in the morning. A cup of tea is my first hot beverage of the day (well, I am English) and, thankfully for this book, there is also a lot of interesting physics at play here, too. All tea comes from the tropical plant known as *Camellia sinensis*, which grows best in a warm climate with long days, cool nights and an abundance of rainfall. The varying types of tea such as green, black and oolong are due to how the leaves are processed after harvest – black tea being fermented while green tea isn't, for example. According to legend, tea has been brewed for centuries, beginning in China in around 2,700 BC, but it took thousands of years before it became a popular drink in the country. To make tea, 'kettles', typically made from bronze or iron, were used to boil water. Then in the nineteenth century, copper became more prevalent, given that it conducts heat more efficiently. Later that century the copper teapot, similar to how it looks today, made its way into homes.

In this speed read of tea-making history, the first electric kettle was also being developed around the end of the nineteenth century. In 1891, the US-based Carpenter Electrical Company released an electric teapot with a heating element separate from the water compartment, meaning it took some ten minutes to boil water. Improvements to electric kettles came in subsequent decades, which included the commonsense idea of placing the immersion heating element into the actual water container. Yet traditional steam kettles continued to be popular, especially with the advent of gas-cooker tops in the early twentieth century. While electric kettles are now commonplace in Europe, in the US the steam kettle is still widely used. This is due to cultural preference as well as the fact that the mains voltage in the US is 110–120 volts (compared to 230–240 volts in the UK), meaning it can take just as long to boil water in an electric kettle as it can in a traditional steam one. Once the water is boiled, an electric kettle automatically switches off thanks to a channel within the kettle, typically inside the handle, that carries steam to a thermostat near the base, which when heated to near 100°C will trip the power off. Those using a steam kettle, on the other hand, will be made aware that the water has boiled due to the characteristic noise of a kettle whistle. This is a cylindrical duct placed at the end of the spout, which includes two circular plates that are closely spaced apart inside (see figure on page 30). Both plates have a hole in the middle that allows the steam to pass through.

Despite this whistling noise having been heard for well over a century, nobody fully understood the physical mechanism behind it until 2013 when acoustic engineers Anurag Agarwal and Ross Henrywood, from Cambridge University in the UK, tackled the problem. Agarwal first became interested in the whistling kettle when doing a PhD in acoustics in the US. He discovered that the phenomenon was first tackled by the nineteenth-century British physicist and mathematician John William Strutt (known more widely as Lord Rayleigh).* In Rayleigh's 1877 book *The Theory of Sound*, he compared the mechanism to how birds produce birdsong. But even the great physicist admitted that 'much remains obscure' when it came to how the sound is produced. 'Lord Rayleigh didn't have microphones and similar equipment back then, which would have made it difficult to study, so we thought we would make the measurements and validate his theory,' Agarwal told me.

They found that Rayleigh's theory didn't quite apply to whistling kettles. To investigate further, Agarwal and Henrywood tested a series of whistles of different lengths by forcing air through them at various speeds.[1] The pair found that once the water is near boiling, the steam going through the kettle's spout produces sound at a single, fixed frequency.† When they investigated this surprising result, they discovered that the noise is generated in the same way as when you gently blow over the open neck of a wine bottle.

---

* Lord Rayleigh was awarded the 1904 Nobel Prize in Physics for his work on gases and the discovery of argon.

† Frequency is the number of events, or cycles, per second. It is measured in hertz.

This creates something called a Helmholtz resonator, which in the case of a wine bottle causes sound to radiate from the neck of the bottle at a fixed frequency. In a similar way, the air inside the whistle reverberates like the air in the neck of a wine bottle, producing a characteristic hum at a constant, single frequency.

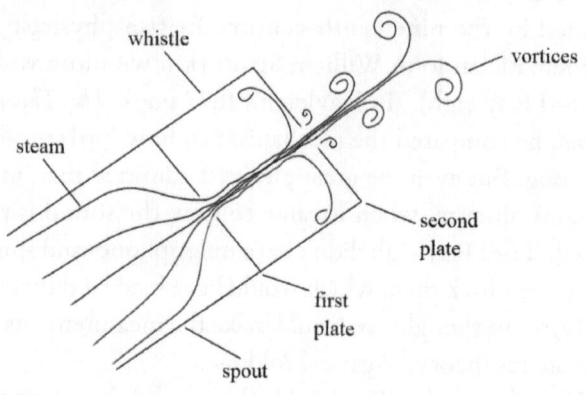

A kettle whistle features two plates that each have a small hole. The fast-moving steam entering the whistle's first hole forms a jet, while the second hole acts to produce mini vortices that are responsible for the characteristic whistling noise.

Once the water in the kettle is on a rolling boil, however, steam is pumping out and travelling much faster. This is when another sound – the whistling we are all accustomed to – kicks in. As the steam in the spout enters the first hole of the whistle, it contracts into a fast-flowing stream of steam. This jet of steam is unstable and starts to break up as it makes its way through the whistle's cavity to the next plate, producing sound waves in the duct between the plates. By the time it gets to the second plate the steam jet hits the hole and produces vortices outside the spout. These mini whirlwinds just happen to produce sound at the same frequency as the sound waves in the duct; the note produced being determined by the size and shape of the hole openings and the length of the spout – a bit like a flute. Agarwal found that it is exactly these vortices – a phenomenon called vortex shedding – that causes the sound. The frequency of the sound also increases with the flow rate of the steam, which is why you may hear the sound change the more the water boils. This vortex shedding is the same effect that happens when wind blows over telephone wires, or when the air travels over roof bars on top of your car. Both produce a whistling noise that is not so dissimilar to the physics of the whistling kettle.

Once the kettle is boiled, if you are not in a frantic race to get the kids off to school or head off to work, then you may instead prefer to sit down with a pot of tea. How civilised. The issue for teapot aficionados is pouring the tea while avoiding liquid trickling down the underside of the spout and on to the table. This is known as the 'teapot effect', a term first coined in 1956 by the Israeli physicist Markus Reiner.[2] The phenomenon occurs when the liquid coming out of the spout 'sticks' to the tip and does not flow out cleanly, resulting in some of it trickling down the underside of the spout. It was thought that surface tension and adhesion of the liquid to the surface was behind the effect. Surface tension is the effect you can see on the surface of water that results in a 'film', which allows some insects to walk on water. Deep in a liquid, water molecules are surrounded by other water molecules on all sides, which results in the interactions between the molecules balancing out – giving no net tension. In other words, the molecule pulling above is balanced out by the molecule pulling below, and so on. However, molecules at the very top of the surface do not have neighbours above them – only at the sides or below. 'This results in the molecules bonding more strongly with their neighbouring molecules along the surface, creating a sticky 'surface film', much like the stretchy elastic sheet that pond skater insects use to literally walk on water.

In 1956, however, Reiner discovered that surface tension and adhesion alone were not enough to describe what was going on. Instead, he proposed that when a fluid flows against a surface, it shears. In other words, the part of the liquid that is away from the surface travels faster than the part that is at or near the surface, which is affected more by friction, and that makes it stick to the spout. Pouring quickly, however, helps to avoid this, as the liquid 'detaches' from the surface and flows freely out.

Thirty years later, in 1986, the physicist Joseph Keller, at Stanford University, proposed that the main culprit behind the teapot effect is actually pressure. As the liquid begins to turn out of the teapot, the pressure in the liquid at the pouring lip is lower than atmospheric pressure, since a pressure drop is required to balance the centrifugal forces.* This means that the air 'pushes' the tea against the lip and the outside of the spout, which causes the drip to pour down the underside of the spout. No need for surface tension. In 1989, Keller improved his model to include gravity to explain the

---

* The outward force on a mass on a curved path from a central point – think going on a roundabout in a playground.

point at which it starts to trickle down the underside of the teapot and on to the tablecloth.[*]

Case closed? Not quite. But the end of the dribbling teapot trauma finally came closer than ever in 2021 when researchers led by Bernhard Scheichl from the Vienna University of Technology in Austria declared to the world that they had formulated a 'complete – even though quite technical – theory' of the teapot effect. 'I hadn't actually heard of the teapot effect before,' Scheichl told me. 'But when we looked into it, we realised that no one had really fully explained what was going on.' By filming teapots as they poured water, the researchers discovered that a small liquid drop formed just under the sharp edge of the spout so that the area always remained wet. The size of this drop, however, depends on the speed that the liquid is flowing: if too slow then the drop acts to direct the entire flowing liquid down the underside of the spout, but when poured at a faster rate the water detaches from the drop and pours out.[3] It is this drop, they found, that is behind the teapot effect.[†] The team also carried out a theoretical analysis showing that the effect is an interplay between inertia[‡] of the moving fluid, agreeing with the earlier findings

---

[*] Jean-Marc Vanden-Broeck and Joseph Keller were awarded an Ig Nobel Prize for Physics in 1999 for their work on the fluid flow out of a teapot spout; see Vanden-Broeck, J-M., Keller, J.B., 'Pouring Flows with Separation', *Physics of Fluids A: Fluid Dynamics*, vol. 1 (1989): 156. The Ig Nobels are prizes that have been awarded each year since 1991 to celebrate unusual developments in science, with the stated goal being to 'honour achievements that first make people laugh, and then make people think'.

[†] For a slick video from the team, see youtu.be/jzZ2_Yh8c68.

[‡] Inertia being a property of a body that resists a force that would cause a change in its motion.

by Keller. Other factors include the properties of the fluid itself and the flow of the fluid in a narrow space due to cohesion and adhesion, known as the capillary effect.

That's all very well, but how do you stop it? One possible approach to dribble-free pouring is to coat the inside of the spout with butter. It is hydrophobic, so repels water and gives the liquid a little extra 'push' out of the teapot. Scheichl says, however, there are other ways that thankfully don't involve adding a taste of butter to your tea. One is to use a ceramic teapot, as it is more hydrophobic than, say, glass. The second it to buy a teapot that has a thin, sharp-edged lip on the spout, which allows the water to flow more freely and not get stuck. If it is too curved, then this helps the drop to form underneath the spout. The third has to do with the angle between the spout opening and the neck of the teapot. If it is almost 90° or alternatively very small, then this will also promote the drip. The key is to have an angle somewhere in between, looking somewhat like a 'less-than' sign, i.e. '<'. And a final trick is not to fill the teapot to the brim so that you can pour the liquid out smoothly.

Scheichl and colleagues also concluded from their work that gravity is not a factor in the teapot effect. This only acts to direct the jet of tea and plays no role in its dynamics near the spout. So, if we ever colonise another planet, prospective tea-drinking explorers would still have to put up with stained tablecloths.

If you close your eyes (not now, unless you have an audio-book) and imagine the sound of hot water being poured into a cup, you will probably be able to do so quite effectively. This rather mundane sound has a particular impact on the brain; after all, if you listen to it long enough you might need to go to the toilet. If you ever visit a Moroccan tea house you will see tea being poured from a great height – well over 30cm from the cup – without seemingly splashing the table, or worse, the customer. This pouring technique is done to trap air bubbles in the liquid to produce a layer of foam on top of the drink, which not only adds to the aesthetic appeal of the cup of tea but also to the tasting experience, enhancing its aromas (more on how bubbles can impact a drink's taste in Chapter 10). If you pour water out of a typical plastic bottle, the jet of water that flows happens to be uniform and smooth, known as laminar flow. As you pour, the water hits the surface of the liquid already in the cup and makes little sound. But if you get a teapot and pour water from the same height, you may see that the jet of water breaks up and is louder when it hits the water's surface. This is thanks to the Plateau–Rayleigh instability, in which Lord Rayleigh (him again) showed in the late nineteenth century that a vertically falling column of liquid breaks up into drops if its length exceeds its circumference, or π times its diameter.[*]

In 2023, researchers in South Korea set out to investigate the reasons behind this intriguing discrepancy between noise and liquid break-up.[4] They placed a nozzle with a diameter of a few millimetres about 10cm above a water-filled cylinder. They then sent a jet of water through the nozzle and on to the cylinder and recorded the sounds

---

[*] π being the ratio of a circle's circumference to its diameter, given as the constant 3.14159....

that were made with an underwater microphone. Using a camera, they also imaged the bubbles that were produced as the jet hit the surface of the water. They did this for several different nozzle heights, discovering that the key to making noise wasn't so much the height that the liquid is dropped from, but rather due to the jet stream breaking into droplets. When the jet of water broke up, a louder noise was produced due to more air bubbles being trapped under the water's surface – as happens when pouring tea from a great height (to find out what happens when a single drop falls, see the next chapter). As a smaller-diameter jet breaks up more easily than larger-diameter ones, this explains why pouring from a bottle with a larger opening can be quieter than pouring water from a small-nozzle teapot. Yet the flip side is that when larger-diameter jets do break up, they contain larger drops of liquid that then causes more bubbles and more noise. In any case, they found that if you want to pour your tea as quietly as possible you should do so from a height that is no more than a few centimetres from the surface of the water. Or you could just do it from a great height instead and make your tea pouring a bit more theatrical.

And that is not the only intriguing aspect in the simple process of pouring water into a container. Another is the difference that can be heard between cold and hot water. If you close your eyes while someone pours hot and cold water into separate but identical cups you might be able to tell which is which (try it at home and see). Cold water sounds 'crisper' while hot water is 'duller' and 'splashier'. While the audible contrast between the two has been well documented, an investigation into the physics has not. It was previously thought that vibrations in the container and the water itself were behind the effect, but researchers in China in 2024 were convinced that the bubbles created when the liquid is poured must play a role. To find out they used a beer tower to house water that was either at 10°C (cold) or heated to 90°C (hot). They then opened the spout to pour it into a water-filled container below and used a microphone to measure the frequencies of sound produced, as well as a camera to image the bubbles that were created.

They found that hot water produced more low-frequency sounds, or those with a frequency between 250 and 400Hz. The reason for this, they postulate, is because larger bubbles, with a radius of 5–10mm, are produced, which result in lower-frequency sounds. This is due to the so-called Minnaert resonance, which states that the frequency of sound from a bubble in a liquid is inversely proportional to the bubble's radius; so the larger the bubble the lower the frequency. Pouring cold water tends to create a higher quantity of smaller bubbles with a radius of a couple of millimetres, which create a higher-frequency sound. They attribute these larger bubbles to the fact that hot water has a lower viscosity – the molecules have more energy and so are less likely to 'stick' to other neighbouring molecules that would cause a rise in viscosity. While this difference in viscosity might be hard to spot for water, it is easier seen in honey, as warm honey pours quicker than cold honey because it has a lower viscosity. In this case the higher velocity creates larger air bubbles in the liquid. The audible difference, then, is due to the different size of bubbles produced in the liquid when pouring.[5] Who knew?

Having whetted the appetite with a cup of coffee or tea, it's time for something more substantial. Sitting down for breakfast, you may have some cereals in front of you. Could anything surprising be happening here? Well, yes, otherwise I wouldn't have a full chapter on the physics of breakfast. Pick up the granola and you may discover that all the big chunks of dried fruit or large nuts happen to be at the top. When you reach the bottom of the packet, the last few bowls are mostly just oats with very few chunky nuts or fruit pieces. Try it for yourself – open a box of cereal with different-sized constituents and you will likely find that the transport process has shaken the box and left the larger objects mysteriously at the top, despite them being heavier.

This size-segregation effect was first coined in 1987 as the 'Brazil nut effect' by researchers at Carnegie Mellon University in the US. When a container that has one large ball and several smaller ones is shaken, the large ball rises to the top, even when the larger ball is denser than the others.[6] Given that denser objects sink, buoyancy alone can't solely explain the effect. Another explanation is percolation, in which smaller objects fall into the gaps below the larger one during shaking, with a further reason being granular convection where shaking leads to the grains moving upwards along the walls and then downwards into the middle of the container. The Brazil nut effect proved, er, a tough nut to crack, and another spanner in the works came in 2001 when the inverse Brazil nut effect was discovered. This time larger but lighter objects sank in a shaken bed of smaller grains.[7]

One difficulty in unearthing the reason behind the Brazil nut effect is experimentally studying what happens inside a mixture of objects (rather than only seeing what is going on at the edges of a transparent container). To get around this problem, in 2021 researchers used X-ray vision – no they didn't obtain Superman special powers but used a technique called X-ray Computed Tomography – to track the motions of individual peanuts or Brazil nuts in a box as it was shaken back and forth. The team discovered that the Brazil nuts that were initially lying horizontally adopted a more vertical orientation when shaken. This movement then opened space for the smaller peanuts that were higher in the mixture to drop down, gradually pushing the Brazil nuts all the way to the surface, where they finally returned to lying flat.[8] The key aspect being that the Brazil nuts were only able to move upwards when they were in an upright position. The Brazil nut effect is likely a mix of factors including percolation and convection, and while it can be a great way to sort two different sizes of grain or object in a container, it is slightly annoying for your breakfast cereal and getting that consistent bowl each day. And it is also an issue for the food industry in general, which has discovered various ways to solve the problem – one being to mix the contents on a regular basis. So, if you want to avoid having all the nuts in that first bowl of muesli you could just get a spoon and regularly stir your cereal in the box or container – and get some strange looks in the process.

Perhaps the most famous breakfast cereal from a physics perspective, however, are Cheerios. They are also a favourite cereal for many children, especially the honey or chocolate varieties. Next time you dive into a bowl or prepare one for the kids, why not do a little experiment (if you have time)? Drop a few of the multigrain rings into the milk one after another and watch as they gradually clump together or with the wall of the bowl. This 'Cheerio effect' – where solid particles floating on a liquid are attracted to one other – was first coined in the 1970s by the physicist Jearl Walker from Cleveland State University and has since become a mainstay of fluid mechanics.[9] The reason for the effect is mostly thanks to surface tension (that again). Cheerios are less dense than milk so they want to sit as high as they can on the surface. As the milk is attracted to the sides of the Cheerios they slightly pull the surface of the liquid upwards along the sides of the ring, which forms a concave (think half of a 'u') 'meniscus' (see figure below). The other Cheerios that are nearby then float upwards through buoyancy along the 'curve' of the meniscus – after all, they like to be as high as possible on the surface – forming something akin to a Cheerio elevator. This clumping together due to a meniscus in the liquid is a big deal in fluid mechanics and has even been used to describe how fire ants are able to clump together to make rafts to avoid drowning. At least it gives you something to talk to your kids about when they ask for yet another bowl of honey Cheerios.

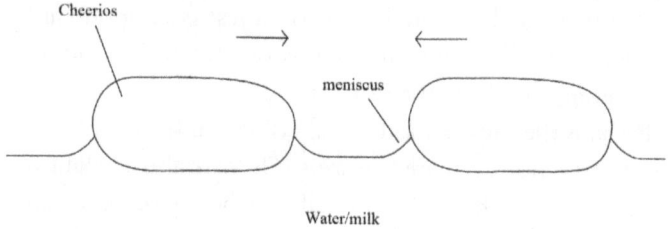

As a Cheerio floats on the water/milk, it creates a concave meniscus, which attracts nearby Cheerios and helps them clump together.

And finally in this physics tour through breakfast, if you really want a nice Sunday-like leisurely morning, then cereal is just the first course, to be followed by a hot breakfast that usually involves an egg of some sort, whether scrambled, poached, fried or boiled.* When you usually heat a liquid, it changes from a liquid to a gas – boiling water to make tea, for example. So, if you put whisked eggs in a pan and add heat, you might expect it to stay liquid or perhaps even evaporate. Instead, the mixture turns into an omelette or scrambled eggs – something that seems solid. What's going on? Are raw eggs a liquid and scrambled eggs a solid? Inside the 0.3mm-thick shell, eggs contain two liquids. The egg yolk constitutes about a third of the volume of the egg and is an oil-water emulsion composed of 16 per cent proteins, 32 per cent lipids and 50 per cent water. The egg white, on the other hand, is a protein suspension, being about 10 per cent proteins with the rest being water. Proteins are like chains: some parts of it are attracted to water and others are repelled by it. Simply put, these proteins are initially folded up and 'slip' over each other to act like a liquid. When heated, however, the proteins begin to unfold (or denature) and link up in a process called coagulation resulting in an elastic, solidified gel of proteins.

In 2021, Nafisa Begam and Frank Schreiber at the University of Tübingen in Germany and colleagues went to some extreme lengths to see what was going on in this process. They used X-rays generated by a powerful machine in

---

* For a range of 'egg-speriments' you can do with eggs, see Bertho, Y., Texier, B.P., and Pauchard, L., 'Egg-speriments: Stretch, Crack, and Spin', *Physics of Fluids*, vol. 34 (2022): 033101.

Hamburg, Germany, known as a synchrotron light source,* to pin down the way in which egg whites solidify when heated. 'The solidification of egg white and yolk upon boiling, also known as gelation, is a rather complex process,' Begam told me. 'And eggs represent a model system to study it.' The team bought chicken eggs from a local supermarket and extracted the egg whites. They then heated the whites on a heating stage to 80°C and fired X-rays at it to determine the structure.[10] They saw that at this temperature the protein network in the egg white had already formed. If the temperature is lower, then the network formation is more gradual and takes longer, as one might expect, but the special thing about eggs is that this slower 'gelation' can also result in different textures and appearance. For example, 'hot spring', or onsen eggs, are cooked at temperatures between 63 to 66°C for about twenty minutes. The result being egg whites that are milky and soft, while the yolk is firm and creamy.

For those that don't have a 0.5km-circumference synchrotron in their back garden to see when their eggs are cooked, what else can you do? The annoying aspect with boiled eggs is that the shell stops you from knowing when it is cooked. I have lost count of the number of times I have taken an egg out of boiling water and started taking the shell off, only to discover that the white is still a gloopy mess. This usually involves putting the egg back in and overcooking it, much to the dismay of my kids, who love a runny yolk. Can a

---

* This involves accelerating electrons in a circular particle accelerator, which results in them giving off X-rays that are then directed towards various 'beamline' experiments.

purely theoretical approach help? In 2000, three Spanish physicists considered a spherical egg* and presumed that it cooks when it is at 70°C or higher. They calculated that an egg taken out of the fridge at 4°C and plunged into boiling water would take thirteen minutes to cook completely, or hard boiled, which the authors say is not far from what is observed.[11] Six years later and physicists from Singapore and Hong Kong calculated the cooking time with eggs that were 'prolate spheroids' shaped – akin to a rugby ball, or an egg for that matter.[12] They also carried out experiments that involved inserting a thin thermocouple, or an electrical thermometer, into the eggs via a 2mm-diameter hole to measure the temperature as the eggs cooked in boiling water. They found that the egg was 'well cooked' when it reached 85°C and this took about 750 seconds, starting with an egg at room temperature – similar to the previous findings.

For those who only contemplate runny yolks, one estimate for their cooking time came all the way back in 1998, thanks to the work of Exeter University physicist Charles Williams. He considered a spherical egg and took it as being cooked when the temperature of the boundary of the white and yolk was 63°C. Using the heat-diffusion equation,† the cooking mostly depended on two parameters: heat conductivity and heat capacity. The heat conductivity relates to how quickly the heat spreads, while the heat capacity is a measure of how much heat is required to warm up the substance by 1°C. Williams found that a medium egg with a mass of 57g taken straight from the fridge took four and

---

* The well-known physics term 'consider a spherical cow' comes to mind.

† Known as a partial differential equation.

half minutes to cook when plunged straight into boiling water, but if stored at room temperature it took only three and a half minutes.[13]

Perhaps the ultimate method, however, came in 2025, thanks to physicists in Italy. The problem with eggs is that the yolk cooks at 65°C while the white cooks at 85°C. The researchers carried out computer simulations that examined the evolution of the cooking at different distances from the centre of the egg.[14] The models suggest that keeping the yolk near its cooking temperature requires alternating the egg between 100°C and at a more tepid temperature. After some experiments in the lab, they came up with a recipe for perfect eggs. The method involves starting with an egg at room temperature and then alternating it every two minutes between a pan of boiling water at 100°C and a pan at 30°C for thirty-two minutes. This 'periodic cooking' results in a yolk that remains at a temperature of about 67°C, giving it a similar consistency to a sous vide yolk. The white, meanwhile, has a consistency between sous vide and soft-boiled. The researchers carried out a chemical analysis of the yolks, finding that this method resulted in them containing more micronutrients compared to eggs that had simply been boiled at 100°C. So, they not only tasted better but also had a higher nutritional value too. Looks like the problem has finally been cracked.

3

LOOKING AFTER YOURSELF

FED AND WATERED, IT'S time to hit the shower, not literally, of course. It's happened to all of us — no, not hitting the shower — but just when you have stepped into the shower and got wet, you remember that once again you forgot to buy shower gel or shampoo and are still using up the last remaining dregs from the bottle. No matter how hard you squeeze the container, it just won't let out those last drops. Some people will shrug their shoulders and try to remember next time they are out shopping. Others will resort to opening the lid and putting a bit of water inside to make the contents runnier, or less viscous, while those that have really had enough might reach for the scissors, cut open the bottle and scoop out the last remaining bits. But this is not only a personal issue. Those last few remaining drops can be a serious environmental problem, as each year thousands of not-quite-empty bottles — usually made from plastics such as polyethylene terephthalate or high-density polyethylene[*] – get thrown away needlessly, ending up in landfill. This not only results in a lot of plastic in the environment but also the leaching of the bottle's chemical contents into the soil.

Getting liquids out of bottles is a common frustration in everyday life. A fundamental aspect of a fluid, as we learned in Chapter 1, is its viscosity. Yet not all viscous liquids are the same and that has to do with their individual make-up. Water-based liquids, such as ketchup, have a high viscosity, but compared to soap it is relatively easier to get out of the bottle (although a ketchup bottle can still require a strong shake). This is due to ketchup's high surface tension (which we touched upon in Chapter 2) and the water molecules sticking together rather than to the plastic container. Liquids like shampoo or detergents, however, are made of differ-ent stuff. They contain clever molecules called surfactants. A surfactant has two important jobs that are carried out by different parts of the molecule. The first is that it connects to water molecules, preventing them from bonding strongly with each other. This reduces the surface tension of the water, allowing it to spread over a greater area. When wash-ing clothes, for example, the surfactant allows the water to seep deeper into the fibres. The second job of the surfactant is that it binds with dirt and grease, which is helpful when removing dirt. The presence of surfactants in shampoo and

---

[*] Refill cartridges or bags are more environmentally friendly than buying plastic bottles. Yet polyethylene terephthalate and high-density polyethylene and some other types of plastic that are used for bottles are generally recyclable.

shower gels leads to a lower surface tension, so you might think that this would help. The problem is that these organic molecules tend to stick to the sides or surfaces of the plastic bottle, making it hard to extract the liquid.

In 2016, researchers in the US tackled this sticky problem by designing a coating that led to soap pouring more easily out of a plastic bottle.[1] The team were inspired by the lotus leaf and how its bumpy surface results in liquids, such as water, simply rolling off it (more of the physics of plants in Chapter 5). They coated the inside of the bottles with microscopic 'Y-shaped' structures made from tiny particles of silica – a component of sand. These tiny Y-shaped objects hold the droplets of soap away from the container walls, which creates tiny air pockets underneath the liquid that further helps the container wall to be soap-free. Just as with the lotus leaf, they found that the Y-shaped structures caused the droplets of soap to glide out of the bottle. While the solution was ingenious, it might be a while before we see these structures inside shampoo bottles. After all, the problem of eking out those last couple of drops might not be top of manufacturers' concerns given that including Y-shaped microstructures into bottles would add to the manufacturing process. So, for now, keep storing them upside down, or keep the scissors handy.

Before you get cold and it is time to get out of the bath, or you're running up a large bill using the shower for rather too long trying to get the shampoo out of the bottle, there is one last interesting bit of physics to contend with (I really hope you are not reading or listening to this book in the shower … the bath, however, is perfectly fine) and that is the sound of the dripping tap. It is one of life's little (or rather big) annoyances – the characteristic 'plink … plink' sound that is produced when a single water droplet hits a liquid surface. Believe it or not, the physics of a falling water droplet has been a scientific curiosity for hundreds of years. The first photograph of it occurring was published in 1908 and it took over 100 years before the cause behind it was solved. In 2016, acoustic engineer Anurag Agarwal from Cambridge University (yep, the same person who was behind the whistling teapot in Chapter 2) was staying overnight at a colleague's house in Brazil. Unfortunately, his host had a small leak in the roof of the house and had put a bucket underneath the offending drip in a bedroom to stop the water from going everywhere. The problem being that this dripping noise and the characteristic 'plink, plink' sound kept Agarwal awake for most of the night. 'It was initially annoying, but as soon as I listened to it carefully, I realised that it had a fixed tone,' Agarwal told me.

As Agarwal went downstairs for breakfast the next day, slightly worse for wear, he discussed the leaky sound with his host and another researcher who was staying in the house at the same time. Being curious physicists, they then went about seeing if anyone had studied the effect and came across previous work suggesting that the 'plink' was caused by the impact itself. The result, however, was inconclusive and to their surprise no one had followed up the study. Back in the UK, Agarwal and colleagues began to investigate. When a single drop hits the surface of a liquid, it causes a cavity, or hole, in the liquid that, due to surface tension – that again! – recoils rapidly. Not only does this recoil give rise to a column of liquid shooting out, but also causes a small air bubble to get trapped under the water surface. Using an ultra-high-speed camera, a microphone and a hydrophone, they recorded the sound of the 'plink' both above and in the water.

The Cambridge researchers found that the initial droplet impact is silent, as is the production of the cavity in the liquid and even the jet of liquid. That just left the formation of the bubble. The team discovered that for the 'plink' to be heard, the trapped air bubble must be close to the bottom of the cavity caused by the drop's impact. Yet there was still a puzzle to be solved. If the noise is produced underwater, it wouldn't be so clearly audible. This is because of an 'impedance mismatch' between water and air. In other words, the speed of sound in water is different than it is in air; therefore, sounds produced underwater don't travel well in air. Try it for yourself next time you are in the pool or the bath. Clap your hands or click your fingers loudly and you won't be able to hear much above water. So, something more must be going on for the bubble to be producing the noise that is so easily heard.

When that little bubble is produced, it begins to pulsate at a frequency of about 2 to 8KHz.* Looking further into the problem, they found that the bubble's pulsation drives waves or oscillations in the water's surface itself. This is what makes the drop so efficient to produce an airborne 'plink' – the water surface acts like baffles in a loudspeaker.[2] So, as you lie in the bath and watch a drop of water emerge from the tap by your feet, the sound you hear is not actually produced by the falling drop *per se* but rather the oscillations in the water surface

---

* The frequency being high due to the bubble's small diameter (thanks to the Minnaert resonance, which we came across in the previous chapter).

caused by a small air bubble trapped for a moment underneath the surface. 'The plink happens to be at an annoying frequency, so it grabs our attention just like a newborn baby's cry,' says Agarwal. Intriguingly, the researchers found that changing the surface tension of the water, such as by adding soap, perhaps from a plastic bottle, results in nullifying the 'plink' sound. But not always. It won't do much if the drop is falling from a height, like from the ceiling, as in Agarwal's case. So, if your bath tap never fully closes and it drips, it might be an option to add some bubble bath to save your sanity. Either that or get a plumber to fix the problem.

Nice and clean, it's time to finish off the job and stop hogging the bathroom. One of the most frustrating everyday mysteries is why stainless-steel razor blades blunt so easily and quickly. How is it that the soft little hairs on your face, legs (or elsewhere!), which happen to be some fifty times softer than the steel razor blade, cause such a hard material to eventually fail?* Razor blades are made from 'martensite'† stainless steel with a composition of mostly iron but with 13 per cent chromium and 0.5 per cent carbon. The blade is then coated with a thin polytetrafluoroethylene‡ layer

---

* And it's not just razor blades, but kitchen knives can blunt even after cutting soft materials, such as cheese.

† Named so because the atoms form a specific 'martensite' crystal structure – first observed by the German scientist Adolf Martens around 1890 – that makes it possible to harden through heat treatment.

‡ Better known as Teflon.

to reduce friction between the blade and the skin. While cheaper, lower-quality blades, such as disposables, aren't great to use many times over. Even supposedly higher-quality blades will eventually fail and start pulling on your hairs in an uncomfortable way rather than clean-shaving them off – and that is regardless of how many blades a manufacturer might be able to cram on to a single razor.

There are many explanations for how a razor blade can be damaged as it cuts human hair, such as the edge of the blades become more rounded through wear, or perhaps it's a combination of rust and corrosion. In 2020, Cemal Cem Tasan and colleagues from the Massachusetts Institute of Technology (MIT) in the US wanted to get to the bottom of this blunting conundrum. MIT graduate student Gianluca Roscioli shaved his facial hair with a disposable razor and after every shave examined the blades under a scanning electron microscope to see how the surface of the blade was affected by the hairs. This type of microscope is incredibly powerful and can produce images with a resolution of about 1 nanometre, or 0.000000001m. It works by scanning the surface of an object with a focused beam of electrons, which interact with individual atoms in the sample to produce a signal that contains information about the surface topography and its composition.

While a razor blade might seem smooth to the naked eye, up close it is anything but, being full of tiny imperfections, such as cracks, along the blade's edge. The team found that during shaving there is more going on than just simple wearing down of the metal – or what can happen when a blade is rubbed against something harder than itself. Instead, they discovered that hairs could sometimes open these cracks to form chips, even if a blade is used only once or twice. 'We didn't believe it at first when we saw the cracks even after just a couple of shaving strokes,' Tasan told me. Investigating further, they obtained hairs that had been kindly self-plucked by lab colleagues and used razor blades to cut them (the hairs, that is, not the colleagues) in real time in the electron microscope – effectively carrying out the world's first hair-cutting experiment in such a device. Combining the findings with computer simulations, they discovered three factors at play that produce chips in a razor blade. The first is governed by where the hair impacts the crack: if it strikes the crack just on one side of it, then it will open the crack further and cause chipping. The second is the angle that the hair strikes the blade: if the hair is roughly 90° or perpendicular to the blade, then there is less chance a chip will be produced, but if the hair approaches the blade not head-on but at an angle, there is more chance it will 'push' open the crack, leading to chipping.

The third factor is linked with the material properties of the blade: if the material is softer on one side of the crack than the other, for example, and the hair strikes the softer side, it again could result in pushing open the crack.[3] But this last option also offers a potential fix, at least to stop this kind of chip from occurring from a materials perspective. Tasan and colleagues found that the more homogeneous the blade can be made, the less likely it is that hairs will chip it. In other words, if the surface can be made hard throughout, limiting the softer parts, then it shouldn't chip as much. If you don't want to wait until manufacturers improve their blade quality, however, then there is another option: buy a traditional straight razor, keep it rust free and sharpen it after use with a leather strop. So much for convenience.

Still in the bathroom? What are you, a hormonal teen-ager? You might next trim your fingernails. If you do it every day or most days to keep them looking as good as possible, then you may want to rethink. At least that is according to Cyril Rauch and Mohammed Cherkaoui-Rbati from the University of Nottingham, UK, who were surprised to find how little work had been done on the physics of nail growth, despite, that is, the Greek philosopher Aristotle apparently being interested in the growth of nails (or specifically horse hooves) back in 350 BC. Human nails, which grow around 0.1–0.2mm per day,[4] are a version of hooves or horns. They are skin cells that contain a protein called keratin, which can 'self-assemble' when cells die providing, in turn, the nail stiffness. The nail consists of several parts – the nail plate, which is what you feel when you touch your nail. Underneath that is soft tissue called the nail bed, which gives the nail its pinkish hue. The 'matrix', no, not the film, is the area at the base of the nail, where new nail cells are produced (the cuticle acts as a protective layer for the matrix). The white crescent at the nail base is the visible part of the matrix, called the lunula. The white part of the nail plate that you trim is called the free margin.

A crucial aspect of the nail and how it grows is the connection that it has to the nail bed via adhesive molecules. These are funny-looking things that hold on to the nail and then, as it grows, they tilt forward and stretch before breaking. They then revert back and reattach to another part of the nail that is coming through. This allows the nail to slide forwards and grow in a 'ratchet-like' fashion by continuously binding and unbinding to the nail bed. By modelling the nail's movement using plate mechanics, Rauch and

Cherkaoui-Rbati[5] showed that when the nail grows too quickly or slowly, or the number of adhesive structures change, it causes a 'residual' stress across the entire nail that results in the nail changing shape over time. The problem with residual stresses is that they remain even when the cause of the stress has been removed. According to their simulations, constant trimming, or bad trimming especially at the edges of the nail, can lead to a residual stress, resulting in a change in the nail's curvature and how symmetrical it is over time. This can end up causing conditions such as ingrown nails, which affects millions of people worldwide, as well as pincer and spoon-shaped nails. Indeed, it was thought that these three nail conditions were medically unrelated, but the work shows that they are all related by a single factor: residual stress. While residual stress can occur in any fingernail or toenail, the model shows that it is greater for nails that are larger in size and with a smaller curvature at the edges, which explains why ingrown toenails predominantly affect the big toe.

If you really must trim your nails daily – perhaps you are a hand or foot model – Rauch told me that the best way to stop imparting residual stresses on the nail is to cut straight across it and not at the edges.

As anyone who has brushed long hair knows (unfortunately I don't have that problem), knots are a nightmare. There can be about 100,000 individual hairs on a scalp and the usual way to detangle hair is to work from the tip toward the scalp via short, gentle brushes. Lakshminarayanan Mahadevan from Harvard University faced this exact problem when he was brushing his 5-year-old daughter's hair many years ago. Mahadevan admits that he lost patience when doing so but was left particularly wounded one day when his daughter told him he wasn't very good at the task. The failure, and those comments, stayed with him for decades, until he decided to use mathematics to get to the bottom of it once and for all. Maybe he could also find something out that would help future parents.

Rather than use a full head of hair, Mahadevan and colleagues simplified the problem to just two strands that were both free at one end and clamped in place at the other (the end in this case being the 'scalp'). The strands were braided or entwined like a helix – looking like the famous structure of DNA. They then modelled and carried out experiments using a single stiff 'tine' like a single bristle of a brush that pulls down the helix, disentangling it until the two strands are free of each other.[6] They experimentally measured the force of combing this two-stranded helix and then used it to model a tine's movement as it attempted to unravel the helix. As the tine moves through the helix, tangling increases and this is related to a mathematical quantity known as a 'link density'. The higher the link density, the more tangled the hair is and the greater the pain on the scalp becomes as it gets tugged by the tine. The model showed that starting near the tip end and working your way to the scalp via short brushes – exactly as employed by careful parents – helps to avoid creating a high link density and represents the best and most pain-free way to go about disentangling hair. The model also suggests that the curlier the hair, the shorter the brush strokes need to be, while for straighter hair, longer strokes can be employed without the link density increasing too much and it yanking at the scalp.

Using these mathematical rules, the work even inspired a robot – effectively a robotic arm with a soft brush on the end – that is capable of combing hair. The brush contains sensors to determine the force that is being applied to the hair and then uses the principles discovered by Mahadevan's work to determine how to brush. The robot includes a camera to assess the curliness of the hair beforehand and to stop brushing when the hair is finally disentangled.[7] So far, the robot, which has been built in collaboration with scientists at MIT, has only been set loose on wigs of different colour and curliness, but the scientists think that future iterations of the device could be used in care homes to free up time for carers to do other tasks. However, a few issues need to be, er, combed out before it can be used on people. Indeed, brushing very curly hair[*] is one particular challenge, as is considering a person's level of pain, which as we all know can differ from person to person and, as any parent is well aware, can be especially low for children.

For those with very curly hair, it's unlikely a robot with a brush will be able to help. Indeed, brushing curly or kinky (no, not that sort of kinky) hair is a particular issue for Black

---

[*] Why do humans have such differing hair types? Hair is an evolutionary adaptation to protect the head from damaging radiation from the sun and to protect the body from the cold. Scientists think that hair acts as a barrier to decrease heat loss from the body (in this case, the scalp) to its surroundings. Curlier hair, which doesn't lie flat, allows the scalp to 'breathe' better while still protecting it from the sun. See Lasisi, T., Smallcombe, J.W. Kenny, W.L., et al., 'Human Scalp Hair as a Thermoregulatory Adaption', *Proceedings of the National Academy of Sciences*, vol. 120, no. 4 (2023): e2301760120.

women, who often find it difficult to choose the right kind of haircare product. That was an issue faced by materials scientist Michelle Gaines from Spelman University in the US. She is an African American with curly hair who previously used a chemical relaxer – which works by breaking the covalent bonds* in the hair fibres – to straighten her hair. When Gaines became pregnant, she stopped using them but failed to find something that could do the same job and was more natural.† 'That is when I got into the physics of hair,' Gaines told me. 'Just from doing my hair all the time, it felt like an experiment – changing a haircare product or changing a combing technique – and I thought this would be a good area to study.'

The common hair classification system includes four categories: straight (type 1), wavy (type 2), curly (type 3) and kinky (type 4). In the 1990s, the celebrity hair stylist Andre Walker expanded this system to create 'the hair chart' – no points for originality there – that today is currently the gold standard for classifying hair. Each of the four categories received three subtypes: 'a', 'b' and 'c'. 1a, for example, is straight (fine), while 2c is wavy (loose waves) and 4c is kinky-coily (tight coil).‡ Yet the system still suffers from

---

* This is a type of chemical bond that involves the sharing of electrons to form electron pairs between atoms.

† A longitudinal study by researchers at Boston University's Black Women's Health Study found that the long-term use of chemical hair relaxers by postmenopausal Black women was associated with an increased risk of uterine cancer; see Bertrand, K.A., Delp, L., Coogan, P.F., et al., 'Hair Relaxer Use and Risk of Uterine Cancer in the Black Women's Health Study', *Environmental Research*, vol. 239 (2023): 117228.

‡ See www.curlcentric.com/hair-typing-system.

ambiguity, as it relies on simply inspecting the hair in question. Gaines and colleagues set out to see if a more numerical approach could put the hair chart on a firmer footing. To do so, they collected samples of different hair styles that were naturally shed from the scalp following brushing or combing. Using microscopes and cameras, they determined the diameter, cross section and even the three-dimensional shape of the strands. They then measured the force and stress of the strands as they were first uncurled – the so-called 'uncurling force'. Wavy hair required very little force to uncurl, but curly hair required much more force. When they then stretched the fibres until breaking point to determine the tensile strength, they found that type 4, or kinky hair, was more brittle as compared to straight hair. 'This means that the more you do to manipulate it, even combing the hair, the more susceptible it is to breakage,' says Gaines.

The team came up with a 'stretch ratio' – a parameter that defines the force required to uncurl a strand until it is straight. The ratio was zero for straight hair (since it can't be uncurled), 0.8 for wavy hair, 1.1 for kinky and 1.4 for curly.[8] They found that this stretch ratio scales with the number of complete waves of curls or coils – or how many contours – exist in a certain length. Within 3cm of hair length, for example, wavy hair has one, curly has two, and kinky has three or four. Gaines says you can use this to categorise your own hair by plucking out a strand and placing it unstretched on a ruler, then count how many curls there are within 3cm. Gaines next wanted to delve a little deeper into what makes curly hair, well, so curly. The hair shaft comprises multiple layers of cells. The inner cells make up the cortex, which consists of intertwined bundles mainly made of the protein keratin and air pockets that gives hair its mechanical properties, such as its strength. The cortex also contains the pigment melanin, which determines hair colour and is also the principal absorber of solar radiation in hair (and skin). The outer translucent layer is called the cuticle and it consists of six to ten rows of flat cells that overlap each other, a bit like roof tiles.[9] These cells open when exposed to water and are sealed, protected and hydrated when shampoo and conditioner is applied.

Using an atomic force microscope to examine the varying hair types, Gaines found that the cuticles are very different for various types of hair. For wavy hair, the cells in the cuticles are larger and more spaced apart with edges that are smooth, while for curly and coily hair they are closer together. She also found that the cuticle is thicker for straighter hair with six layers all around the fibre. For curlier hair, however, there tends to be six on one side and two on the other.* This not only helps to explain some of the mechanical properties to allow the curling to happen but could also explain why curly locks dry out faster than wavy or straighter hair – it being harder to keep water locked inside due to a lack of cuticle layers. Given that most consumer products cater mostly to either straight or wavy hair, Gaines hopes that more work into the mechanical and structural properties of hair will help design better-tailored consumer products for all hair types and not just those focused on a select few. 'It's still a surprise to me how little work has been carried out by the big cosmetic companies on

---

* See A., Syed, *Curly Hair: Structure, Properties and Care* (Oak Brook, IL: Hasnia Publishing, 2023).

curly hair,' says Gaines. 'One explanation could be that most of the scientists that are doing this work don't have this kind of hair, but I think there is a lot more we could be doing.'

For those with straighter hair, once you have brushed it, the next step might be to put it in a ponytail. You might think that physics can't possibly have anything to say about the humble ponytail, but that hasn't stopped physicists from wondering why a bundle of hair fibres seem to have a distinctive bob-like shape? A ponytail contains thousands of individual hairs and as gravity pulls a bundle of hair downwards from the hair tie, the interaction between the many individual threads causes the tail of the ponytail to protrude outwards like a fan. This gives the ponytail body and volume, and the curlier, or springier, the hair, the more it swells outwards. In 2010, scientists at the consumer-goods giant Unilever contacted physicist Raymond Goldstein from the University of Cambridge and colleagues to see whether statistical physics could say anything about such hair bundles. 'During our first meeting I suggested finding the shape of a ponytail as a first step toward understanding more general problems like tangling,' Goldstein told me. 'We felt that there should be such an "equation of state" for a ponytail.'

The researchers investigated by first obtaining human hair 'switches' – a type of commercially available hairpiece – and measured the curvature or curliness of a sample of individual hairs. They then developed a model that included the stiffness and curliness of individual hairs in a ponytail as well as its weight and length. Using this they managed to formulate a 'ponytail shape equation', which could predict how the shape of a ponytail varies as it increases in length from the hair tie.[10] From the computed shapes they found that the so-called 'launch angle' of the ponytail from the hair tie, or the angle at which the outermost hairs emerge from the vertical (imagine a vertical line straight up the middle of the ponytail and the angle between it and the outermost hair), is remarkably consistent between ponytails of different lengths, being about 17°.

The theory also determined whether, for a certain length of ponytail, it ends up looking like an upside-down fan or whether it instead initially fans out but then arcs back and becomes nearly vertical at the bottom, as happens for longer ponytails. In other words, a long ponytail with flatter hair hangs down as gravity overrides the springiness of the hair, while a short ponytail made from springy hair tends to fan outwards against the downward pull of gravity. 'There is a general connection between the "pressure" with which a bundle pushes outward – from the waviness of interacting hairs – and the shape of the bundle itself,' notes Goldstein.

The work was not purely for academic purposes, however, with Goldstein adding that it allows companies like Unilever to quantify the effects of their hair-care products to describe aspects such as 'body' and 'volume' that otherwise might be considered 'vague and subjective'. So, when you next roll your eyes at that shampoo advertisement claiming to boost your hair's volume (because you're worth it), there might be some supporting theoretical physics behind it.

Another interesting aspect about ponytails – yes, there's more – is when out jogging. You may have noticed that when a ponytailed jogger is running, the ponytail usually swings from side to side. This is despite the jogger's head and body moving up and down when running. To find out why this happens, Joseph Keller from Stanford University (who studied the dribbling teapot of Chapter 2) modelled the ponytail as a pendulum that hangs and swings from a support base – in this case the head. To solve what impact this might have on the ponytail, Keller turned to the work of US astronomer George William Hill. In the late 1880s, Hill was trying to figure out if the periodic motion of the Moon around the Earth was stable – quite an important topic. This is related to the dynamics of the Sun–Earth–Moon system, what is known as a 'three-body problem' in physics, and is tricky to solve. Through the endeavour, Hill derived in 1886 what is now known as Hill's equation.* It says that if there is a 1:2 relationship in frequency between a pendulum and what it is fixed to, then any deviation in the motion of the pendulum, no matter how small, will grow exponentially over time. So, if the pendulum and what it is fixed to are both moving up and down at a frequency ratio of 1:2, if the pendulum has a slight deviation to the side, say, then the amplitude of that motion will grow until it reaches a stable value.

Back to the runner with a ponytail. The average runner will have a rhythm of about a 140 to 160 steps each minute, or about 2.7 cycles per second, meaning that their head will bob up and down at this frequency. The average 25cm-long ponytail has a natural frequency of about 1.4 cycles per second – or roughly half the natural cycle of the head. As there is a near 1:2 relationship between the frequency of

---

* This being a second-order linear ordinary differential equation.

the runner's bobbing head and the ponytail, then according to Hill's equations it means that even a small amount of side-to-side motion – such as what can be provided when a runner takes alternative steps – will grow until the ponytail is really swaying from side to side. And this is exactly what happens if you watch someone with a ponytail start running.[11] And you also don't need a ponytail to see the effect, or even need to be running. It also happens if you are wearing a hooded sweatshirt or jacket with tassels hanging down from the collar. All you need to do is walk at a brisk enough pace and watch as the tassels start to swish from side to side with increasing amplitude until they are swaying forcefully. Who would have guessed that a runner's oscillating ponytail could have a connection with trying to figure out if the Moon's motion around the Earth is stable!*

Finally, all clean and hair sorted, it's time to get dressed. Many people just throw something on and don't worry too much about it (and who hasn't seen photos of themselves from five or even ten years ago and noticed they still wear the same T-shirt? Thanks, Facebook). Others, however, like to avoid mainstream trends. These so-called 'hipsters' are individuals who are known to actively dress in unique ways as a statement that they are not part of the majority. The saying goes that 'hipsters avoid labels and being labelled'.

---

* In 2012 Joseph Keller, Raymond Goldstein, Patrick Warren, and Robin Ball were awarded the Ig Nobel Prize for physics for 'calculating the balance of forces that shape and move the hair in a human ponytail'.

But do they? In 2014, mathematician Jonathan Touboul from Brandeis University in the US was not specifically seeking to answer this question directly but was instead interested in the surprising finding that random fluctuations (akin to 'noise') in neurons could result in synchronicity and order. Using techniques from statistical physics, he wanted to look at this effect using the simplest model possible and settled on studying conformist and anti-conformist 'individuals' that could choose between two states (denoted in the model as +1 or -1).[12] The model considered random fluctuations between these states as well as interactions between them. 'Originally, I did not think of it as modelling hipsters,' Touboul told me, stressing that he did not directly study the hipster/fashion aspect *per se* in the work. 'But after being quite exposed to hipster culture where I live, I slowly realised that the phenomenon could – half-jokingly – be qualified as the hipster effect.'

The model considered the time needed for a state to become mainstream. Information can vary in speed, some people may follow fashion and see a trend emerge quickly, while others rely on seeing it in real life or by word of mouth. Touboul, who admits to 'loving' the hipster culture, found that the population of non-conformists, or hipsters, initially act randomly but then, in statistical physics speak, undergo a phase transition where members synchronise with each other in opposing the mainstream. Basically, when extrapolating the findings to hipster culture, those people who want stand out from the crowd end up synchronising with each other to develop a shared look – going against the very nature of what it means to be a hipster. A further surprise emerged when there were equal populations of conformists and hipsters. In this case, the entire population tends to switch randomly between different trends. For example, if most individuals in the conformist camp shave their beard, then most hipsters will want to grow a beard, and if this trend propagates to most of the hipster population, it will lead to a new, synchronised switch to shaving to get away from the trend. 'What was surprising was how few ingredients the model needed to generate such a rich repertoire of behaviours,' says Touboul, adding that he can't offer any tips on how a hipster can become, well, a true hipster. 'Fashion or styles are really not my domain,' he adds.

Yet there is a rather funny story attached to this work to end the chapter on. The website MIT Technology Review reported on Touboul's research when it was first published in 2019 and illustrated the article with a stock image from Getty images. The picture showed a man in a checked shirt and beanie with the caption, 'Shot of a handsome young man in trendy winter attire against a wooden background'.* Nothing unusual going on so far. Yet the website then, in the words of MIT Technology Review's editor-in-chief Gideon Lichfield, received a 'furious e-mail' from a man who complained that he was the person in the photo. MIT Technology Review was even threatened with legal action because they used the photo without his permission as well as slandered him, presumably because they implied he was a hipster.† But after some digging from MIT Technology Review it turned out that the image – of an archetypal

---

* See www.technologyreview.com/2019/02/28/136854/the-hipster-effect-why-anti-conformists-always-end-up-looking-the-same.

† See www.theregister.com/2019/03/06/hipsters_all_look_the_same_fact.

hipster – was actually of someone else and had been properly licensed from an image library. 'He'd misidentified himself,' noted Lichfield.* 'All of which just proves the story we ran: Hipsters look so much alike that they can't even tell themselves apart from each other.'

It seems the research touched a nerve.

---

* See twitter.com/glichfield/status/1103040764794363904.

4

CREATURE COMFORTS

GROWING UP, I HAD a dog called Jasper (I say 'I', but it was really my parents who looked after him). As I was so young, I can't remember too much about Jasper, who was a black mongrel, but the photos of me when I was a toddler show we were best buddies. As a teen I then grew up with a grumble* of five pugs, who each had their own personality, and despite being so small, shared an incredible ability to slobber all over everyone. The connection between child and dog is a special one, and the bond people have with their pets can be so strong that many owners consider them a part of the family — even more than some particular (human) members. According to UK Pet Food,† in the mid to late 2010s just under half of the UK population had a pet. But the COVID-19 pandemic really kick-started a pet-owning

---

* The collective noun for group of pugs is a 'grumble'. For anyone who knows their character, this seems fitting.

† Formerly known as the Pet Food Manufacturers' Association.

revolution, with the number increasing as people sought a companion to help them through the dark days of lock-downs. In 2022, the level of ownership was as high as 62 per cent of the UK population, consisting of about 38 million pets – more than half of which were dogs and cats (12 and 11 million, respectively).[1]

A survey of almost 6,000 people carried out in 2020 by the University of York in the UK found that pets provided a source of considerable support during COVID-19 lockdowns, with 90 per cent of respondents saying their animal helped them cope emotionally, while 96 per cent noted they helped them to keep fit and active.[2] Despite coming with significant costs (food, toys, veterinary bills and your own time look-ing after them), having a pet brings many benefits, not least meeting new friends when walking the dog, higher levels of exercise or just having a companion at home to help you relax and unwind.* Research in 2019 found that dog owner-ship was associated with a 31 per cent lower risk of death by heart attack or stroke compared to non-owners. Those who had suffered a heart attack and were living alone with a dog, meanwhile, had a 33 per cent lower risk of death compared to heart-attack survivors living alone but without a dog.[3]

While I may be more of a dog person than a cat person, let's also give some love to our feline friends, as studies

---

* As always with these kinds of studies, there can be conflicting findings. Work in 2024, for example, found that while there were physical benefits of having a dog, the mental health rewards were less clear cut. See Parsons, C.E., Landberger, C., Purves, K.L., et al. 'No Beneficial Associations Between Loving with a Pet and Mental Health Outcomes During the COVID-19 Pandemic in a Large UK Longitudinal Sample', *Mental Health & Prevention*, vol. 35 (2024): 200354.

have found that cats can help to lower stress and make their owners feel happier, more confident and even sleep better.[*] And a cat can have pluses for kids, too. A survey of more than 2,200 children aged between 11 and 15 found that those with a strong bond with their cat had a higher quality of life, being more fit, attentive, and less lonely.[4] And the benefits are not just one way, of course, with domestication offering safety, food and care for animals.

As anyone with a pet knows, a key part of the morning routine involves taking care of them, whether going for a quick walk to do their business/get some exercise or making sure your cat, dog or fish are fed and watered (not so much for the fish in this case). But while you might rush this part of the day in order to get to your desk or school drop-off, it is worth taking a moment to consider some interesting physics at play as you walk or feed your pets.

Let's start with giving them something to drink. As members of a group of placental mammals that specialise in eating flesh – called carnivora – mature cats and dogs have 'incomplete' cheeks. This brings advantages, as it allows them to open their mouths as wide as possible to deliver blows to their foes or to catch prey (and we know that cats are particularly good at that). But when it comes to drinking it has some disadvantages. It means that they can't

---

* Some research shows that living with and having a close bond with an animal companion does not necessarily lead to significant mental health improvements in people with a serious mental illness, such as bipolar disorder. See Shoesmith, E., Lorimer, B., Peckham, E., et al., 'The Influence of Animal Ownership on Mental Health for People with Severe Mental Illness: Findings from a UK Population Cohort Study', *Human–Animal Interaction* (2023) doi.org/10.1079/hai.2023.0027.

fully close, or seal, their mouth to form a vacuum (except, that is, when weaning), which makes suction impossible. If they tried to suck, they would just bring air into the cheeks. Conversely, humans do have complete cheeks, allowing us to drink by creating negative pressure, sucking water or other fluids through a straw into our mouths and down our throats. Dogs and cats, therefore, need another way to get liquid – usually on the floor in a bowl or in the form of a muddy puddle – into their mouths in an energy-efficient manner.

Putting a bowl of water down for a thirsty dog in the morning can result in a huge amount of mess as the liquid is sloshed and spilled around, particularly if there is more than one dog having a go at the bowl. A cat, on the other hand, seems to lap it up elegantly without too much collateral to mop up afterwards. As your pet drinks there is a lot going on that even our eyes are not quick enough to capture. All we see is a cat or dog seemingly 'licking' the water or milk into their mouths. But is this technique efficient? Try it for yourself, just maybe not in front of anybody, and you will soon see it can be hard work. One might think that, instead, cats and dogs basically curve their tongues and 'scoop' the water, a bit like cupping your hands together and scooping up water from a stream. While it was once thought that the mechanism cats and dogs use to drink was much different, over the past decade it has been revealed that they use similar drinking strategies that differ in a couple of interesting and nuanced ways.

The old showbusiness saying advises never to work with children or animals, and it seems a similar case could be made when it comes to research. At least that is what Sunghwan Jung from Cornell University, who at the time was working at the Virginia Polytechnic Institute and State University (now Virginia Tech), and colleagues found when they tried to study cats drinking. The first issue was getting someone to agree to lend their cat for research purposes. They finally got around the issue when Jung's colleague – Roman Stocker from the US Massachusetts Institute of Technology – volunteered the services of his cat, Cutta Cutta. The second challenge was then getting the cat to drink in front of a camera. It turned out she wasn't very thirsty; after all, cats can get their water from other sources, such as food.

Thankfully, the researchers had a trick up their sleeves and squirted a bit of tuna juice into the water to entice Cutta Cutta to the bowl.[5] When they finally managed to record some laps with the high-speed camera, they saw that the feline extended her tongue to the surface of the liquid but did not penetrate it. Instead, the bottom of the tongue rested on the surface. The liquid then 'stuck' to the underside of the tongue, and as the tongue retracted it brought up a column of liquid underneath at a speed of about 1m/s (metre per second). After a split second, gravity pulled the water back down into the bowl, but before this could happen the cat slammed its mouth shut over the top of the column. This cycle occurred some four times per second, each time the cat capturing about 0.1ml of liquid, or less than a teaspoon.

But what about dogs? Previous research suggested that dogs drink differently; after all, they are quite good at spilling water all over the floor. It was thought that dogs instead curl their tongue back into a ladle and scoop the water up directly into their mouths (domestic cats slightly curl their tongues as they drink as well but not as much as dogs). Yet in 2011, researchers at Harvard University used high-speed video combined with X-ray footage to show that while some water was scooped up by the curled-back tongue, it mostly rolled off before it could get into their chops. They concluded that dogs perhaps took the same approach as their arch-rivals when lapping up liquids by allowing water to adhere to the tips of their tongues.[6]

To finally get an answer for how our canine friends drink, Jung and colleagues took nineteen dogs of different breeds (not literally, of course, but with their owners' permission) and filmed them as they drank. Getting the dogs to drink was easier than in the case of Cutta Cutta, but the researchers still asked the owners to exercise the dogs beforehand to make sure the mutts were thirsty. They then filmed the dogs drinking water using two cameras – one placed in the bowl itself looking upward into the dog's mouth and another to record the dog as it drank from the side.* They saw that as a dog drinks its tongue does indeed curl backwards into a 'ladle' shape, but the difference with a cat is that the dog instead breaks the surface of the liquid with its tongues as it drinks, creating a splash. Like a cat, the water adheres to the underside of the tongue and then as the dog retracts it into its mouth, a vertical column of water is created that the dog snaps its jaws around. Yet the dog's retracting tongue is quicker than a cat's – with a speed more than 1m/s – and it can drink more per lap than a cat. One–nil to dogs.

---

* For a video of a dog drinking, see youtu.be/63Ch2pNkZwU.

To examine dog drinking further, the team created a model of a glass tube to simulate the dog's tongue and mimicked the acceleration and the creation of a water column in the fluid. They found that the water column produced by a retracting tongue pinches off at a certain point and detaches due to gravity, but that a dog is smart enough to close its jaws just before the column breaks, allowing it to catch as much water as possible. Who said that dogs are stupid![7] The larger the dog, the more of their tongue they put into the water, meaning that bigger dogs produce a bigger splash, but Saint Bernard owners probably already know this. 'Domestic cats are gentle drinkers, they don't want to make a mess,' Jung told me. 'But dogs don't really care about splashing around.'

So why are cats so careful? One explanation Jung offers is that they perhaps don't want to get their sensitive whiskers wet. One interesting aside from our discussion of domesticated animals is that Jung and colleagues also filmed lions and tigers drinking (from a safe distance). This was done at a zoo, not in the wild, and still required adding some chicken blood into the water to make it more enticing. Surprisingly, they found that the big cats splashed liquid much more than domestic cats and also curved their tongues in a manner more similar to dogs than domestic cats, showing there is more to animal drinking than meets the eye, or rather the high-speed camera. Indeed, Jung and colleagues wanted to go further with their study of dog lapping. Their plan was to use a technique called Positron Emission Particle Tracking, which involves introducing a radioactive particle, known as a tracer, into the water that can then be imaged as it moves through the dog's mouth. Yet the university didn't approve the work as it would be too dangerous for the dogs.

'You need to be careful when doing research on people's pets,' adds Jung.

Fed and watered, it's time to take the dog out for a walk. Many animals – including humans – have evolved to be good at walking long distances at a steady pace. When mammals walk, it is a repetitive and precise movement that is incredibly efficient. The body's centre of mass – an imaginary point at which the whole mass of an object is concentrated – is located just below the navel. As humans walk, we do so under what is called an inverted pendulum mechanism with the centre of mass oscillating between two points in space. You can do this experiment yourself and put your finger just below your navel and then take a few steps. As you stand two feet together, the centre of mass is at its highest point, but when you take a stride, it dips slightly relative to its height from the floor, rising again as the other foot passes the forward-planted foot. If you traced the location of the centre of mass over a few steps, it would look like that of an (upside-down) pendulum.

This inverted pendulum has been documented in many different animals and that includes quadrupeds.* The reason why this simple mechanism is found so much in nature is because it is so efficient. The efficiency can be explained in terms of mechanical energy, or the conversion of kinetic potential energy as the person moves into a potential gravitational

---

* In the case of quadrupeds, the forequarters and hindquarters in dogs behave like two independent bipeds.

energy change as the centre of mass moves up and down. The amount of energy recovered depends on the walking speed, but can be as high as 65 per cent in adults trotting along at 4.5km/h per hour, and even higher in some dogs at about 70 per cent.[8] This goes some way to explain why some breeds can walk such long distances without seemingly much effort.

Cats, however, are not built for walking long distances, and in 2008 researchers in the US studied six domesticated cats to find out why. They were filmed as they walked on 'force plates', which is basically a set of large scales that measure ground-reaction forces in three dimensions, similar in some sense to a Nintendo Wii Fit Balance Board (remember that?). They found that when the cats walked normally, their energy recovery was only about 40 per cent – much lower than humans and dogs.* When the cats moved in a stealthy way, like when they were hunting that unfortunate bird or mouse, they 'walked' incredibly slowly, shoulders up, body close to the ground, so that each step was placed as quietly as possible. In this case the efficiency dropped to 20 per cent or lower.[9] The careful movement has everything to do with stability and caution but is terrible for efficiency. In other words, cats have traded efficiency for stealth.

Back to dogs. Going for a walk with your mutt is not just about getting out, but also to have some fun through play. This usually involves throwing a ball countless times, which the dog happily catches and brings back with such

---

* A similar efficiency to toddlers who are just learning to walk (35 per cent); see Banks, M., *The Secret Science of Baby: The Surprising Physics of Creating a Human, from Conception to Birth – and Beyond* (Dallas, TX: BenBella Books, 2022).

enthusiasm, like they have never done it before. Some breeds are particularly keen on the ball–chase game: running after it at top speed, catching it in their mouths (well, sometimes) and then turning incredibly quickly to run back, just to do it all again and again and again. Turning movements at speed, like catching a ball to then run back to the owner, require balancing many forces, such as the rotating body as it turns and the changing centre of mass of the animal relative to the floor, as well as maintaining sufficient friction with the surface to avoid slipping over. The question is how do some dogs manage such quick turns with spindly legs without sliding over or, worse, breaking one or more of their legs?

A possible answer came in 2022 when Christofer Clemente from the pleasantly named University of the Sunshine Coast in Queensland, Australia, and colleagues analysed turning ability and the trade-off between running fast and running safely (with cornering being an example of when turning too fast can be dangerous). They first looked at quolls* but then turned, pun intended, to what makes dogs good at turning. They teamed up with a group of Queensland dog trainers who are particularly adept at making their agility dogs run around corners quickly. The team used high-speed cameras and drones to film quick-turning movements in twenty domestic dogs of varying sizes. They made the dogs run around an oval-shaped course, which contained markers where the dogs had to turn – a bit like an obstacle course you might see at the dog show Crufts. It was designed such that the dogs could reach top speed on a

---

* If, like me, you didn't know what a quoll is, it is a carnivorous marsupial native to Australia and New Guinea.

straight, but they would have to manage that speed through a tight turn. It would be reasonable to think that the larger the dog, the more difficult it is to turn, but it, er, turns out to be a bit more nuanced than that. Medium-sized dogs (think Border collie or those that weigh between 15 and 20kg) could manage turning at top speed better than smaller dogs (like a papillon) or larger dogs (such as a German shepherd).* The turning ability from medium-sized dogs was put down (sorry, wrong turn of phrase) to their longer front legs and shorter, stiffer back legs.[10] Clemente says the finding came as a bit of a surprise as he initially thought that small dogs would be the best turners, but perhaps, he adds, breeding has played its role in making them less able to turn at speed. 'If I could, I would do the study again but with cats,' Clemente told me. 'But they would probably all just run away.'

Many animals, such as lizards, cats and squirrels, use their tails to move and balance. Cheetahs, for example, use their tail to help them perform rapid and exceptionally tight – almost 180° – turns when hunting prey in the wild.[11] But what about dogs? A dog's tail's role in communication – both friendly and unfriendly – is well documented and is a part of the vertebrae that seems to convey a dog's pleasure as it wags furiously from side to side. In 2022, Ardian Jusufi, from the Swiss Federal Laboratories for Materials Science and Technology in Dübendorf, and colleagues examined what role the tail plays, if any, when it comes to mutt movement. To find out, Jusufi teamed up with Katja Söhnel from the University of Jena in Germany to fit Border collies with

---

* It might not be as straightforward as this given that greyhounds, which weigh about 30kg, are the real turning specialists.

reflecting markers to different parts of the dogs' bodies. The dogs were then filmed with high-speed cameras and their movements captured in 3D, which allowed the researchers to build a mathematical model, like a 'virtual dog', as the hounds ran and jumped and moved about. 'Nearly all the dogs had no issues being shaved and having markers attached,' Söhnel told me, 'and in the end, they were very relieved to finally perform the jumps.'

The model comprised seventeen segments to account for the head, neck, body, legs and paws. They then simulated how changes to the motion of the tail could affect movement in general. The models concluded that the tail did not help with balance or when changing direction, and made almost no difference in a dog's trajectory as it jumped in the air.[12] They even found that removing the tail entirely had little effect on movement. This seemingly 'negative' result could, however, lead to more interesting questions. 'Could it be that the effect is so subtle as to not be captured by the model, or that it is used to correct pelvis position rather than whole body posture change?' Jusufi suggested to me. It seems for now, however, that a tail wag is just about communication; in other words, something they use to persuade you to throw that ball once again.

Having a dog means taking them out in all weathers, and that includes the pouring rain. Even if you do a very quick run to the park and back, once inside the house another inevitability occurs – your mutt shakes its fur, almost drenching you in the process as well as everything else in the room. Wet fur is a poor insulator and water droplets tend to stick to a wet animal's fur due to surface tension (that again). Water has a thermal conductivity – a measure of the ability to conduct or transfer heat – of 0.6 Watts per metre per Kelvin, which is twenty-five times greater than that of air, and twelve times greater than that of dry fur. This means that a wet animal loses heat incredibly quickly and if it cannot dry itself rapidly then it faces hypothermia. According to calculations, a wet 27kg dog, with only 0.5kg of water in its fur, would use up a whopping 20 per cent of its daily calories to air-dry itself. If this happened on a regular basis it would be a matter of life and death.

In 2012, David Hu and colleagues from the Georgia Institute of Technology in the US wanted to know how dogs manage to shake the water off so effectively.[13] They investigated by getting out a hosepipe and drenching five breeds of dog – chihuahua, poodle, Siberian husky, Labrador and a chow-chow, as well as a domestic cat, which must have really enjoyed that day – and filming the 'wet dog shake' with a high-speed camera. They found that the dogs and cat generally shook soon after being squirted. A shake began at the head by rotating the head and shoulders and then this 'wave' travelled down the rest of the body. The movement was so powerful that the skin could end up whipping around the body – the loose skin travelling about three times that of the backbone. As the Labradors were so easy to work with, Hu and colleagues took advantage and examined the shake in more detail (thankfully the cat was spared). They taped a drinking straw vertically upright to the Labrador's fur so they could track it mid-shake. They found that the straw moved about 90° on each side as the

Labrador dried itself (imagine vertical at top and then horizontal at the sides of the dog's body) – a movement that generated a force of about fifteen times that of gravity. At such force, the water droplets have no option but to fling off, mostly doing so when the fur changes direction at either side of the 'spin'. They found that this action can remove 70 per cent of the water in just a few seconds.[*] If the dogs had tight skin, then the researchers found that they would only be able to eject about 10 per cent of water, so the loose skin is very important.

The physics problem that animals encounter is that the smaller they are, the harder they have to 'spin' to eject the water in order to achieve a sufficient dryness level. Given their size, the dogs shook with a frequency of about 4 to 7Hz (or four to seven times per second) – with a chihuahua shaking at 7Hz and a husky or Labrador at 5Hz (a domestic cat was almost 10Hz). Indeed, Hu and colleagues studied the shake in many animals, finding a huge range of drying frequencies. Animals such as mice shake at an incredible 30Hz, which results in a force of seventy times that of gravity, so strong they must close their eyes to avoid damaging them. The same physics of a self-drying dog, or cat, applies to a washing machine, which has a usual spin cycle of 100 rotations per minute, or 17Hz.[†] If the machine was small enough to only fit a pair of socks, say, it would need a serious spin cycle to dry the garment, but a bigger drum allows the forces to remain high without requiring a large frequency to get a sufficient dryness. In other words, the

---

[*] For a video of a shake in action, see youtu.be/AFzWJ6P2iyY.

[†] One rotation per minute being equal to 0.0166Hz.

wet-dog shake is nature's analogy to the spin cycle of a washing machine.[*]

The team even built a 'wet-dog simulator', which was lined with fur and could spin at various frequencies to see whether they could mimic real life. They found that the minimum acceleration to remove the smallest drops from the fur was about twelve times Earth's gravity. This lined up perfectly with Hu's findings in which no animal employed a drying force less than that – the kangaroo matched it, but presumably living in Australia they don't get wet that often. Intriguingly, Hu also discovered that domesticated animals tend to shake slower than non-domesticated animals of the same size. The reason for this is unknown but Hu offers a hypothesis: 'People take care of their pets and so these animals might not have the same muscular strength anymore. It seems they have lost some of that ability.' Blame those pooch parlours, then.

The wet-dog shake is well adapted to removing water, but not so much for other liquids. That is why animals cannot rid themselves of oil if caught up in an oil spill: the oil has a lower surface tension than water, meaning it seeps deeper into the fur or feathers and doesn't evaporate, so is much harder to expel. Yet fur itself does have some self-cleaning properties. If you take your dog out for a walk and it gets covered in mud, by the time it gets home dry it may be generally clean (on its own without needing to jump into a lake). Where did the mud go? Perhaps all over the car, but in 2023,

---

[*] Apart from removing water, the move is also used by animals to eliminate irritants, such as tangles and parasites, particularly in areas on the back and neck that are unreachable by self-grooming or licking.

researchers in the US discovered another possible mechanism. They obtained individual clean hairs, both animal and synthetic, and then placed them in a flow of distilled water that was rich in titanium dioxide – a material that is 'wildly' sticky – that acted as a fouling agent. In one experiment they fixed both ends of the hair so that it didn't move, finding that furs from semi-aquatic animals such as beavers and sea otters captured the least amount of titanium dioxide, followed by terrestrial animal furs (coyote and springbok) and then synthetic fibres, which got covered in the stuff.[14] Yet when they fixed only one end, they discovered that the more the hair flexed or bent, the better they all performed.[15] Longer hairs – as they bent more – were cleaner compared to short hairs. The researchers suspect that the hair's ability to bend causes the adhesion between the dirt and fur to be broken, giving it an anti-fouling property. They think that air probably acts in a similar manner to water – the mud literally falls off as a dog starts running.

Given the discussion so far has been focused on dogs (well, I did say I was a dog person), let's move on to some interesting physics that is unique to cats. Indeed, one aspect of a cat's behaviour that is couched in physics is how they always manage to land on their feet. Believe it or not, this is a problem that has occupied scientists for hundreds of years. Even the great physicist James Clerk Maxwell, who managed the not-so-small feat of unifying electricity and magnetism, thought about why and how this can happen. It is so fascinating because cats can completely change their orientation during a fall. Imagine yourself falling – say you have just done a big jump on a trampoline and are just at the top of the jump and on the way back down – do you think you would be able to easily rotate your whole body so quickly as you fall? Perhaps not, but cats can easily rotate their bodies in a way that they always fall on their legs – even if they are held upside down and let go from a height (in the interests of animal welfare, don't try this at home).

This cat trick was first captured on film in the late 1880s by the French scientist Étienne-Jules Marey. He pioneered the development of photography and created a new technique called chronophotography that was able to capture a series of still images quickly. In 1882, he made a chronophotographic gun that could take twelve consecutive frames a second, with all the frames recorded on the same picture, allowing him to photograph movements in detail that had never been seen before with the naked eye. Marey wanted to use this new contraption to study animal motion and in 1894 trained his device on a falling cat. The experiment simply involved someone holding a cat upside down by its feet from a height and letting it go as it was filmed turning over in the air. The study immediately threw out the idea that the cat was somehow using the surface it falls from (such as the hand) as a fulcrum point to begin the turning motion. The pictures instead showed that the cat had no rotational motion at the start of its descent and somehow acquired angular momentum while in free fall. Marey concluded that the cat manages such a feat by using 'the inertia of its own mass to right itself'.[16]

While the images clearly showed what was happening, it took a few more decades before anyone could explain the physics behind it. In 1969, engineer Thomas Kane, from Stanford University in the US, and colleagues modelled the falling cat by breaking its body (not literally) into two separate, but connected, rotating parts. The model consisted of two identical cylinders, representing the front and rear of the cat, that were joined by a special 'no twist' joint, which could bend (think of it like two tin cans attached by a bit of wire ... and don't tell me that physicists never make accurate models).[17] When they simulated the ways in which this theoretical cat could move, they saw similar motion to what is seen in real life.

What moggies do is all about the conservation of momentum and the impact of changing the moment of inertia – a measure of how the body resists being rotated. Initially the cat is, say, upside down, but as it falls it brings its front legs into its body, while stretching out its back legs. This changes the moment of inertia between the front and back of the body. By putting its legs closer to its body, the cat is lowering its moment of inertia and this results, by the conservation of momentum, in a higher spinning speed. In other words, the front of the body turns quicker. This is the same physics that happens when figure skaters outstretch their arms when spinning before quickly pulling them back into their body to rotate at incredible revolutions. Back to the cat. As its back legs are stretched out, this has the opposite effect, and the spin is slower (and in the opposite direction to the front part of the body). With the cat's front body now in position to have a soft landing (on its legs), it then reverses the effect and brings its back legs in and extends its front legs, which has the same effect on the back end of the body by rotating it quickly but keeping the front end rotating slowly by extending the front legs. Still with me? If it sounds complicated, it is, but it is also incredibly effective. It only takes a second for the cat to complete the motion to make sure it lands on all four feet.

As well as providing an explanation for how cats always land on their feet, the work showed that it was possible to rotate without the help of an external force. This brought with it other applications. The US space agency NASA became intrigued by Kane's work and supported him to study techniques that astronauts could use when moving around in a weightless environment. Even a professional gymnast was brought in to test the moves on a trampoline. The techniques that cats use to right themselves are now taught to astronauts to maintain their orientation in space, which kind of begs the question what happens when cats go into space – do they have perfect control of their bodies? Well, not quite. Videos* of cats in micro-gravity conditions show them completely disorientated by the effect as they continuously try to right themselves while floating around.

Better not send a colony of moggies to Mars, then.

Pets, of course, are not only about cats and dogs, so let's end the chapter on fish, which after all can be the first pet for a child, just to make sure they can keep something alive before moving on to a rabbit, guinea pig, cat or dog. If you happen to have a classic spherical goldfish bowl, then you might have gone up to feed your fish or fishes only to not

---

* For a video, see youtu.be/O9XtK6R1QAk.

immediately be able to spot one. Indeed, a spherical bowl has a knack of somehow hiding the little blighters. There is a cool physics explanation for why this happens, and it is not because your goldfish has discovered a way to camouflage itself against a background. Instead, it is all to do with the weird effect that light has when it goes from one medium (water) into another (air).

The physics is explained by Snell's law, named after the sixteenth-century Dutch astronomer and mathematician Willebrord Snellius who discovered it in 1621. It describes the change of angle that happens when waves like light pass through a boundary.* It's the same effect that occurs when you put your hand in water and find that, by simply looking at it, your actual hand is not exactly in the same place as it seems. When light travels from a material with a higher refractive index to one with a lower one (as happens when light goes from water to air), some fun things happen. At a certain angle, the so-called critical angle,† the light doesn't pass through but rather travels across the interface – the surface of the glass, for example. For angles greater than the critical angle, the light again doesn't pass through but also doesn't go along the interface. Instead, it reflects back into the medium – a phenomenon called total internal reflection.

What does all this have to do with a fishbowl? Well, in 2009 physicists in China calculated how a goldfish might

---

* This is known as the refractive index of a material – a dimensionless number that indicates how much light bends or changes direction in the medium. For air at 0°C and one atmospheric pressure, the refractive index is 1.000293 and for water it is 1.33333.

† The critical angle from water to air is 48.6°.

be able to 'disappear' in certain positions for an observer (i.e., you).[18] Using Snell's law, they found that if the goldfish is located in the centre of the bowl it will easily be spotted, regardless of where the observer is (the reason being that the light from the fish strikes the bowl at small angles everywhere). If the fish now moves towards the glass bowl, things can change.* If you are standing right in front of the fish then you will be able to see it, of course, but if you start moving away horizontally (therefore creating an angle between you and the fish), then at certain angles the fish will 'disappear' due to total internal reflection. The physicists calculate there is a certain space in which if your eyes are located then the fish would 'disappear' (see figure below). You could try it for yourself: when the fish is near the edge of the bowl, start moving around and see where that blind spot of total internal reflection occurs.

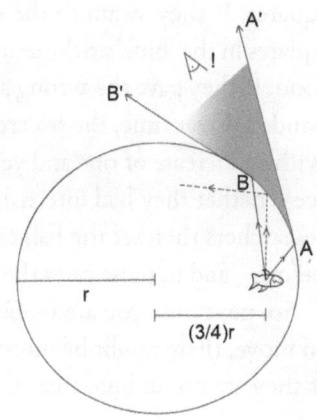

A fish located near the edge of a goldfish bowl will be unobservable to someone located within the shaded region. This is because the light rays from the fish between A and B will be totally internally reflected (dashed line) and not observable for someone located within A' and B'. Adapted from Zhu, Y and Shi, F., 'Why Does the Goldfish Disappear in the Fishbowl?', The Physics Teacher, vol. 47 (2009): 424–426.

---

* They found this happens when the fish is located about three-quarters of the radius from the centre of the tank.

Some might think that fish can be a bit of a boring pet, just swimming about all the time from one end of a tank to the other. Yet a common myth is that goldfish have a memory of only a few seconds, something that was reinforced by the character Dory in the hit film *Finding Nemo*.* 'I suffer from short-term memory loss,' says Dory to the clownfish Nemo. 'It's true, I forget things almost instantly. It runs in my family. Or at least I think it does. Er, umm, where are they?' Despite fish being on the receiving end of such a bad rap, they are capable of some incredible mental tricks. It has long been known that some fish, such as angelfish and guppies, can count, at least to four, but researchers in Germany in 2022 revealed they could do even more. They showed cichlids (and stingrays) a collection of geometric shapes, such as four squares. If these objects were coloured blue, it meant 'add one', while yellow meant 'subtract one'. The animals were then shown two new pictures: one with five squares and one with three squares. If they swam to the correct picture (i.e., to the five squares in the 'blue' arithmetic task) they were rewarded with food. If they gave the wrong answer, they went away empty-handed. Over time, the sea creatures learned to associate blue with an increase of one and yellow with a decrease of one. To see whether they had internalised this mathematical rule, the researchers then set the fish calculations they had never seen before – and in most cases the fish gave the correct answer.[19]

So, next time you are tapping on the fish tank for your pet to move, there might be more to them than meets the eye … if they are not hiding, that is, due to the laws of physics.

---

* It is thought that goldfish have memories that last weeks, months or even years. This is based on research that was conducted decades ago, so why the myth persists is a bit of a mystery.

THE GREAT OUTDOORS

DURING THE COVID-19 PANDEMIC lockdowns, many people not only turned to their pets for comfort but their gardens, too. It was a place of sanctuary or somewhere to entertain the kids and get them out of the house for a few hours. The first lockdown of 2020 coincided with the start of spring in the UK and just when it came into force, the Sun came out and with it longer and warmer days. We flung open the patio doors so that we could all breathe a little easier rather than being on top of each other in the house. We were lucky enough to have an outdoor space, but others, sadly, weren't – a hardship made worse by the closure of parks during the first few months of the pandemic. Yet we were not alone in such an endeavour in getting some fresh air. An online survey carried out in 2021 by the University of Surrey found that during the first year of the pandemic, some 88 per cent of people stated that their gardens helped to relieve stress and anxiety, while 78 per cent of respondents noted that their outdoor space gave them a better appreciation of nature. Going outside during the third national lockdown, which began in the UK in the New Year of 2021, however, was not quite as attractive, given the cold, dark days of January and February. Yet we still did, and despite years of reseeding, the lawn never quite recovered from all that wet trampling.

A fondness for the outdoors is especially so in the UK, where the temperate climate generally results in cool, wet winters and warm, damp summers, which provide good growing conditions for plants. Especially in the summer months, some part of the day will be spent outdoors. Indeed, gardening is one of the most popular pastimes that is enjoyed by all ages – even my kids appreciate watching the strawberries or tomatoes grow in our small vegetable patch, and they particularly enjoy eating them, even if they are still somewhat green. In a separate survey by the University of Surrey in 2021, some 43 per cent of those polled said they spent most of their time in the garden, well ... gardening, with the next most popular activities being resting, sitting and lying down (27 per cent); reading (21 per cent); watching and feeding nature (14 per cent); listening to music, the radio and podcasts (13 per cent); and, perhaps surprisingly for the UK, enjoying the weather (11 per cent).[1] Those with gardens also like to spend money on it. According to estimates, a whopping £18.6 billion was spent on UK gardens in 2021, with the average total spend per adult with a garden being £690. The over 65s spent the most (that pension must go somewhere, I guess) on 'garden goods' such as plants and tools, accounting for an average spend of £3.80 per week.[2]

Next time you sit in the garden, perhaps with a freshly brewed coffee in hand after taking the dog for a walk, close your eyes. You may hear the birds singing, the rustling of leaves in the wind or the sound of insects buzzing. All this flora and fauna on offer not only brings with it great joy, but – and you probably know where I am going with this by now – a lot of interesting physics, too.

Rain has a special place in the British psyche, especially when it comes to predicting when the next downpour will occur. It can be annoying when the heavens open and you have things to do outside, such as going for a run, hanging the washing out or even doing some gardening. Yet rain's power to change the environment has fascinated humankind for generations. After all, drops of rain can bring relief to a region hit by drought; cause flooding, if it is heavy enough; or indicate a change of seasons. Rainfall begins in clouds where 'proto droplets' are formed. These droplets merge and when one becomes too heavy to be suspended in the clouds, it falls to Earth, accelerating under the force of gravity. As the drops fall, they interact with the air and perhaps collide with each other to form natural rain as we see it on the ground.

If you have forgotten to water some flowers for a couple of days during a dry spell, then you will likely see the plant droop gradually, with the flaccid stems unable to support the weight of the flowers. Yet add some water, either thanks to a downpour or with the garden hose, and they will, hopefully, straighten up and revive. Plants have evolved a wide variety of movements, from the slow growth of shoots (on the scale of a millimetre per hour or less) to the ballistic firing of seeds and spores at 10m/s from some types of fern. A plant can be thought of as being about 75 per cent liquid and 25 per cent solid (holding a similar ratio to that of a human newborn baby). Plant cells are surrounded by a stiff cell wall, which prevents them from using soft contractile proteins to deform and generate movement like 'muscles'. Instead, plants use water to 'move', and to do so they transport water in and out of their cells via a process called osmosis – the movement or diffusion of molecules across a semipermeable membrane. This process changes the volume of the cells and the tissue stiffness, allowing the plants to grow or reposition themselves.[3]

All plant stems respond to stimuli like light and gravity – called tropisms. Phototropism is the ability to grow towards a source of light, while heliotropism, which is demonstrated by sunflowers, is the ability to track the position of the Sun during the day.[*] This movement allows plants to capture more of the Sun's rays to convert into chemical energy through photosynthesis. Phototropism was first reported in the early nineteenth century when the Italian scientist Sebastiano Poggioli described the effects of red and violet light on the leaves of the sensitive plant (*Mimosa pudica*). It was later found that cress (*Lepidium sativum*) would grow vertically in the dark, but when placed in a box in a room with a window would sprout towards it. Charles Darwin and his son,[†] Francis, provided a detailed description of photosynthesis in their 1880 book *The Power of Movement in Plants*.

---

[*] Phototropism and heliotropism seem to have a different mechanism; see Brooks, C.J., Atamian, H.S. and Harmer, S.L., 'Multiple Light Signaling Pathways Control Solar Tracking in Sunflowers', *PLoS Biology*, vol. 21, no. 10 (2023): e3002344.

[†] Charles Darwin had ten children, although three died in childhood.

One experiment involved the duo placing a plant in a dark room that was lit on one side by a candle. They discovered that the plant would bend its stem towards the light.

Yet the optics of how phototropism worked remained a mystery until 2023 when researchers in Switzerland examined a mutant version of thale cress (*Arabidopsis thaliana*). This small plant grows abundantly in cracks in pavements in Europe and Asia, and its small white flower is also dubbed 'the lab rat of the plant world' due to its simple genetics and ease of use. The researchers examined a mutant version of the cress that, surprisingly, had transparent stems. They found these plants failed to respond to light compared to the wild-type version. In the normal thale cress, air is contained in channels between its cells, but in the mutant version this part is enclosed with water instead. The team concluded that the air in the channels acts as a light gradient to help the plants 'read' where the light is coming from.[4] While humans and other animals have light receptors only in our eyes, plants are covered in them, which allows them to 'see' from all sides to steer towards light. It's a bit like we humans having eyeballs all over our bodies, which I am sure I have seen in a *Resident Evil* boss before.

Plant stems usually oppose gravity, but not always. Another intriguing aspect about plants is why some stems grow up against gravity while others seem to sag under its effect. If you sit in your garden, you might see different stem geometries. For example, the shoots of thale cress produce vertically upward-growing stems while Himalayan balsam (*Impatiens glandulifera*) has a stem that oscillates between upwards and downward growth before growing vertical. Snow-in-summer (*Cerastium tomentosum*), meanwhile, has a stem that sprouts vertically downwards before growing upwards, with the white willow (*Salix alba*) having stems that grow vertically downwards. But what generates these different shapes?

Lakshminarayanan Mahadevan from Harvard University (of hair-brushing fame in Chapter 3) has also been fascinated with the physical mechanisms that allow plants to 'move' – whether it is the growth of leaves or the opening of flower buds (more on that later). In 2017, Mahadevan and colleagues went about answering the taxing issue of how different stem shapes could arise from a mathematical perspective. Stems sense and respond to their environment, but their growth is constrained by gravity and physical properties of the stem, such as its mass and elasticity. The researchers developed a model that used just two parameters: a sensitivity to growth, or how strongly the plant senses its own surroundings; and the elasticity of the shoots.[5] They found that the interplay between these two parameters was enough to recreate the diverse stem shapes that are seen in nature. They found that a stem that goes straight upright against gravity requires a high sensitivity of growth and a low stem elasticity, while an oscillating stem – one that goes both with and against gravity – needs a high sensitivity of growth and a medium elasticity. The shoots of the weeping willow, for instance, initially grow upwards towards light, but then as the shoots are soft, they sag under the weight of gravity, falling towards the ground. It is this parameter space of sensitivity and elasticity that plants seem to explore to produce a range of different types that are seen in nature.

Once the stem has emerged, whether up or down or both, leaves generally follow. Leaves are mostly two dimensional in that they have one dimension (thickness) that is much smaller than the other two – width and length. While the genetic and biomolecular mechanisms that give rise to complex three-dimensional shapes has been studied, the biomechanical role was less clear. Indeed, a big mystery is how leaves can produce different types of geometries such as saddle, helix and ripples. The plantain lily (*Hosta lancifolia*), for example, features saddle-shaped leaves that are rippled along the edges. In 2009, Mahadevan teamed up with Haiyi Liang from Harvard University to recreate the lily's leaf shape in the lab. To do so, the duo used flat, foam ribbons measuring approximately 11cm × 4cm (about the size of a small bookmark). They then stretched this artificial leaf at the edges beyond its so-called elastic limit – the point after which the object becomes permanently deformed and doesn't go back to its original shape. When they then let go, allowing the foam ribbon to relax, they saw that it naturally formed a saddle-like shape.[6]

When the researchers applied even more strain, basically stretching it four times harder than they did before, the object not only relaxed into a saddle shape but also developed ripples, or small undulating waves, along the edges, exactly as seen in the lily. An analogy to this effect is when thin potato slices are dropped in hot oil – you get a bulging, smooth middle but ripples forming on the edge of the crisp. The thinking is that as an elongated leaf grows, certain cells lengthen more than others, and this difference in growth creates a strain. In other words, as the cells elongate more along the edges compared to the middle, it induces a strain, which bends the leaf as well as creates a serrated surface that you sometimes see on large leaves.

While the mechanism for this so-called differential growth is dictated by the plant's genes, work in 2018 found that strain is also enough to account for more than just saddle and rippled leaves. Researchers in the US and Singapore collected live leaf specimens from several species and dissected them. They found that doing so on one side affected the shape of the other side, which suggested the presence of a residual stress when the leaf is in its natural state. Using a computer model they then applied strain across the leaves and when they tweaked the parameters could reproduce four types of shape that can be found not only in leaves but also in orchid petals: twisting, helical twisting, saddle bending, and edge waving.[7] They even used the theory and replicated the shapes in a hydrogel a few centimetres long in which the strain was represented instead by the flow of oxygen to control the stiffening of the gel.

As this work showed, strain is not only responsible for the shape of long leaves. The real star of the show in any garden are the flowers – and one of the most spectacular instances of plant movement is the delicate unfurling of a flower bud. In 2011, Liang and Mahadevan were back for more plant action and this time studied the Oriental lily (*Lilium 'Casa Blanca'*). This is a classic oriental lily with sweetly fragrant, pure-white flowers that curve delicately back towards the end of each petal, revealing dark-red stamens – the pollen-producing reproductive part of the flower. A lily bud consists of three inner petals surrounded by three outer sepals – the greener part of the flower that protect it. When it is ready to emerge, over a few hours to a few days, the petals and sepals invert gradually and then peel open like a banana to form a blossom. This timescale suggests that growth is likely the reason for the bloom, and one reason for the petal movements could be the difference in the growth rate between the two surfaces of the petal. If certain cells elongate more than others, then this causes a strain that can bend the tissue – as we saw in the case of the leaf.

To test this theory, the researchers placed an Asiatic lily's stem in water, and in an environment where the humidity and temperature were kept the same.[8] The plant was under continuous fluorescent lighting and was filmed with time-lapse video to record the blooming process at intervals of a minute. The pair found that over four and a half days, the Oriental lily's young buds – being about 10cm long and 2.5cm wide – turned white, but they also absorbed about 200ml of water and increased in length by 10 per cent and in diameter by 20 per cent. The inner petals began to wrinkle, while the outer petals initially remained smooth but wrinkled towards the end of the bloom. When that happened, the petals reversed curvature and bent outwards. This wrinkling was a clue that the edges may have been growing quicker than the smooth middle. In a separate experiment, the team removed the edges of the lily's petals, finding that what remained did not curl back as much. They then developed a mathematical model, which showed that strain on the edge, like what happens with the growth of long leaves, not only helps the petals to open but also helps to produce the necessary curl.

You can thank mechanical strain induced by the differential growth of plant cells, in part, for producing the spectacular range of flower blooms and leaf forms that can be seen in countless gardens, including your very own.

Once a plant is in full bloom or at a mature point in its life-cycle then reproduction, the next critical process, begins. For plants, this can involve spreading pollen with the help of pollinators such as bees, where, in return for some nectar, bees (more on bees later in the chapter) get their fur covered in pollen, which they deposit elsewhere.* Some plants have developed other mechanisms to spread their seeds, which can involve a whole host of tricks, and one used by ferns that belong to a class called leptosporangiate are couched in physics – or, rather, ballistics. They have evolved one of the fastest known catapults in the plant kingdom. When you think about a medieval catapult, a particular design comes to mind. This includes three important aspects: a way of storing energy in a 'spring'; a trigger to release all that pent-up energy; and finally, a break to stop the arm at a specific point to let the projectile fly. Some ferns, however, have managed to merge all these aspects into one specific structure called an annulus.

The underside of such fern leaves contain small brown blobs. These feature hundreds of so-called sporangium that are about 0.3mm long. They feature a small stalk with a 'ball' on the end that hold spores (see figure opposite). Covering half the outer surface of the ball are twelve to thirteen special cells in a line: the annulus. The catapult motion begins when the plant dries out. The sporangium loses water to evaporation, and this causes it to contract, which increases the tension in the cell. As this happens, the

---

* It is believed that pollinators such as insects and hummingbirds build up positive charge when flying. Negatively charged pollen is then attracted by the electric field formed by the pollinator and the plant.

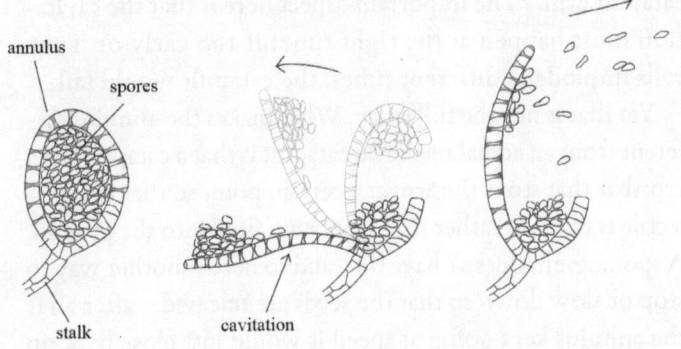

A fern's sporangium uses its twelve-cell annulus as a catapult (left). The annulus bends backwards until cavitation occurs (middle), which propels the annulus forwards, tossing the spores great distances (right).

sporangium gradually opens as the annulus bends backwards, breaking away from the other cells on the other half. This movement, which takes tens of seconds, is a bit like forming an 'O' with your thumb and forefinger before bending your finger back but keeping your thumb still (yet you won't be able to bend your finger the whole way back unless you want to break it). This is the first aspect compared to our medieval catapult, with the bending motion storing a lot of elastic energy. As the annulus continues to bend backwards, the strain in the cell increases rapidly and at a certain point the negative pressure in the cell is huge, about 90 bar.* At that point, the individual cells in the annulus implode. This rapid cavitation releases the elastic energy and results in the annulus shooting forward like a

---

* In comparison a car tyre is at about 2 bar.

catapult arm.* The important aspect here is that the cavitation must happen at the right time: if too early or if the cells implode at different times, the catapult would fail.

Yet that is not the full story. What makes the annulus different from an actual medieval catapult is that a catapult has a crossbar that stops the arm at a certain point so that the projectile is released rather than just being fired into the ground. A sporangium doesn't have this, and so needs another way to stop or slow down so that the seeds are released – after all if the annulus kept going at speed it would just close back up again with the seeds still inside, or perhaps they would just roll out on to the ground beneath. To avoid this, the fern has another trick up its, er, annulus. When the annulus is shooting forward – a process that takes only microseconds – it reaches about 40 per cent of the way back to its original position before deaccelerating abruptly. This is due to the movement of water back into the cells, which results in something called 'viscous drag'. This slows the moving annulus and the rapid deceleration mid-course is enough to throw out the spores with an initial velocity of about 10m/s – equivalent to an acceleration of about 100,000 times the acceleration due to gravity.[9] As water continues to flood in, the annulus continues to move slowly back to its original position. It's incredible to think that nature has managed to create a catapult out of just twelve cells lined up in a row – all thanks to the movement of water in and out of the cells.

Another example of physics-based seed dispersal will be familiar to anyone with a lawn. While many gardeners try desperately to rid their garden of weeds, one of the most

---

* Micro being 10^{-6} or 0.000001.

invasive weeds, or herbs as some would call it, is the humble dandelion (*Taraxacum officinale*). Spotted in many parts of the world, it features yellow flowers that turn into a head of white seeds, known as a dandelion clock. The seeds are easily dispersed by a child kicking a dandelion head over the lawn or by picking it and simply blowing on it. Indeed, it only takes a short puff of breath to get the seeds airborne and if there is no child on hand to help then a gust of wind will do. Most seeds land a few tens of centimetres away from the plant, but in certain conditions, such as warmer, windier environments, a dandelion seed can float through the air for more than a kilometre before falling to the ground, all thanks to some cool propulsion.

An individual dandelion seed head contains a small stalk, known as a beak, which at the top has around a hundred bristles called a pappus, and at the bottom is the 'fruit'. The pappus looks somewhat like the head of a chimney sweeper's brush (or imagine a stick attached to the middle of a bike wheel that only has spokes and no rim or tyre). If you were to get such a brush or rim-less wheel and hold it above your head while jumping off a building, it wouldn't do you much good, but things are rather different for a lightweight pappus. The mechanism for how it works was only discovered in 2018 when researchers in the UK examined the aerodynamics of its flight.[10] They put individual dandelion seeds in a vertical wind tunnel – basically a fan at the bottom of a tube – while a fog machine was used to image the air inside the tunnel to see what kind of patterns were created as the seed 'flew'. The individual bristles on the pappus are so thin that they only take up 10 per cent of the total area (imagine the bristles lying in a disc), the rest is literally thin air. Despite this, calculations have shown that the bristles can generate four times the lift that would be generated by a solid disc of the same radius. Of course, the bristles are not lying perfectly flat like a disc, but instead are arranged in such a way that when the pappus falls, air flows between them. When this happens, the researchers found that the air forms circulating rings, like smoke rings or vortexes, above the pappus head. Intriguingly, the ring is not attached to the pappus, but above it, floating free. It is this mixture of size, mass, shape and how the air flows through the individual filaments that generates the vortex ring, which generates lift.

The researchers found that the porosity of the filaments seem perfectly tuned to stabilise the vortex to keep it above the pappus. A year later, a separate team calculated the airflow around the pappus,[11] modelling it as a collection of rods that are arranged like the spokes of a bicycle wheel (again, no rim). They were able to recreate the experimentally measured wake – the vortex rings – above the pappus, confirming that they provided the seed with stable lift. When they played around by changing the number of 'spokes' in a pappus, they found that the optimal number for flight was about 100 filaments – exactly what is found in your garden. So, when you next see a dandelion clock, don't be annoyed that it is yet another weed that needs removing from the lawn but rather marvel at how nature has managed to create such wild solutions to a tricky problem like seed dispersal.

A garden would not be complete without some wildlife. Most will have birds hunting for worms in the ground or a fox hunting for your leftovers in the bin at night. Perhaps that corner of the garden you hardly touch is infested with snails, spiders or woodlice. In summer, flying insects are everywhere, from wasps trying to spoil your picnic to more welcome visitors, such as butterflies and bees. Summer in the UK also means a lot of rain that, while beneficial for plants could, could, in principle, be a big problem for insects. The raindrops may be small for us, but for an insect like a moth or a butterfly it can be like being hit by falling bowling balls from the sky. Painful. The nearest you can get to feeling this is to watch the 1989 hit film *Honey, I Shrunk the Kids*, in which inventor Wayne Szalinski (played by Rick Moranis) accidently shrinks his four children to 6mm tall. In one scene, the miniature kids are trying to get back to the house via the lawn when the sprinkler is turned on. They must then run a gauntlet of falling raindrops, dodging them like they are giant boulders dropping from the sky.

Thankfully, insects have a way of turning these bowling balls into damage-less ping-pong balls. Many studies had examined the impact of slow-moving droplets on an insect's back. Yet this is quite different from reality, where raindrops are moving at 10m/s, so Sunghwan Jung, from Cornell University (from Chapter 4 on how cats and dogs drink), and colleagues took a more realistic approach. 'We tried to understand what strategies insects use to avoid catastrophic injuries from being hit by raindrops,' Jung told me. To do so, they took high-speed images of fast-moving drops striking moth, butterfly and dragonfly wings to determine how they deal with the droplets. An insect's wing contains a waxy coating that is superhydrophobic (like a leaf) and so repels water. This is a start, but it is not quite enough. Investigating deeper, Jung and colleagues found that the wings also contain an array of tiny bumps or micro spikes that can hold a droplet and help it move around. This array of spikes, they discovered, first helps to spread the water droplet out and then, secondly, puncture it, turning the single drop into tiny beads that roll off the insect's wings.

This insect amour system is so effective that it reduces the contact time between the water and the insect by up to 70 per cent compared to not having this protective layer.[12] This is crucial, as raindrops are colder than the body temperature of the insect, so the superhydrophobic layer combined with the piercing structure also helps an insect to keep its body temperature in flight. 'This special structure is how they survive in the wild,' adds Jung. Indeed, you can see how effective it is if you happen to use a urinal that has a urinal matt installed. If you look at it (not too close), then you will see it is made up of many small spikes. In a similar way to the insect wing, these break up the stream of urine so that it doesn't splash back. So, you can thank moths for keeping your shoes and trousers dry next time you use a public toilet.

When it comes to flying insects, perhaps bees are the most famous. In 1934, the French entomologist Antoine Magnan wrote in his book *Le Vol des Insectes* that when he and his colleague, the mathematician André Sainte-Languë, applied the law of fixed-wing aerodynamics to flapping insect wings, he was unable to explain how they could fly. This so-called 'bumblebee paradox' stated that their wings are too small to get their rather large bodies off the ground. Despite the paradox being explained in the meantime, this thinking even made it into the 2007 *Bee Movie*. 'According to all known laws of aviation, there is no way a bee should be able to fly. Its wings are too small to get its fat little body off the ground,' the narrator declares in the movie's opening sequence. 'The bee, of course, flies anyway, because bees don't care what humans think is impossible.'

Bees can get airborne, of course, and we now have a much better understanding of how they do so – mostly thanks to high-speed cameras and experimental rigs involving mechanical wings. Bees beat their wings about a 170 times per second and it was once proposed that they used a continuous flow of air around their wings to generate lift, similar to aeroplanes.* In a simplistic explanation of how a plane keeps in the sky, the air moves around its wings and the curve of the wings act to create a low-pressure area above and a higher-pressure area below, which provides lift.† Engines then provide thrust, which keeps the air moving over the wings. In the 1990s, however, it was discovered that bees use unique aerodynamic mechanisms to fly,‡ which makes sense given that bees and other insects don't have fixed wings and also don't have engines to keep the air flowing over the wings.

---

* For a super slow-motion video of a bee flying, youtu.be/IcU-i7j0uYs.

† This is a rather simplistic explanation, for a more rigorous account see Babinsky, H., 'How Do Wings Work', *Physics Education*, vol. 38 (2003): 497.

‡ Insects are capable of many sophisticated flying behaviours, such as taking off rapidly, hovering and flying backwards.

The wing motion of a bee is not so much up and down like a bird, but can be forward and back when hovering,[*] up and down when in forward flight and even a mixture of the two. But a bee can also tilt its wings during each flap cycle. In 1996, high-speed video and calculations revealed that as bees flap and rotate their wings, a swirl of air is created at the front tip of the wings that generates extra lift. These are known as 'leading-edge vortices', with researchers discovering that they also occur in other flying insects.[13] It was thought that these vortices alone provided the answer for how bees fly, but in 2017 researchers showed that this wasn't the full story.[14] They proposed instead that the vortices act as a stabiliser, which allow the bees to angle its wing more vertical without stalling. Stalling is what happens when, for example, the angle of attack is too great and there is a sudden breakdown of smooth air over a wing. You can see the effect of stalling if you try flying a plane (remote-controlled, of course) as high as you can by sending the nose upwards. As the plane's wings reach a certain angle from the horizontal, the air can't flow properly over the top of the wing, causing the lift to drop drastically and the plane to come crashing down. That wouldn't be good for bees. Instead, the vortices affect the air in such a way as to let the bee angle its wings more sharply to the sky to improve the air flow over the top of the wing to generate extra lift. This then creates a pressure ratio between the top and the bottom of the wing that provides enough lift for

---

[*] Birds can also hover and flap their wings back and forth, but rarely do so.

the insect to keep aloft.[*] As ever with science, one imagines that this might not be the last you hear about how bees fly. But if there was ever going to be a *Bee Movie 2* they could at least set the record straight.

While bees may be plentiful in the summer, in other seasons other insects come out to play. On a misty autumn morning, you may look out of your kitchen window and see countless spider webs all over the bushes, illuminated by the mist. Spider webs are fascinating structures given they are used by the arachnids to catch prey, lift objects and even propel themselves through the air Spider-Man style. This versatility is thanks to the incredible material that results from the long strands of tubulin proteins that make up spider fibres. Such fibres are strong, flexible and lightweight, and their tensile strength – a measure of the load that a material can bear when being stretched without fracturing – is on a par with steel.[†]

Spider webs occur in different shapes and geometries, but the orb structure is perhaps the most classic, made famous by Charlotte, the barn spider, from *Charlotte's Web* by the US author E.B. White. This type of web generally takes a spider about an hour to make and is made from two main types of silk.[‡] The first is called Dragline, or radial silk, and

---

[*] It is thought that smaller insects use different wing movements than medium- and large-sized insects, as they suffer from increased viscous effects from the movement of the air. For a review see Sun, M., '*Colloquium*: Miniature Insect Flight', *Review of Modern Physics*, vol. 95 (2023): 041001.

[†] Spider dragline silk has a tensile strength of 1.3 GPa, while steel is about 1.6 GPa.

[‡] The more than 50 000 known species of spider can produce up to seven types of silks, which come with varying levels of stickiness.

makes up the radial lines going from the outer edges of the web toward the centre. It is produced by the spider's major ampullate gland and is stiff and supports the frame of the web. The second, and where all the fun lies, is called capture, or spiral silk, and it connects the radial lines in a circular pattern. An orb web spider builds the radial lines first from a central hub and then places the spiral threads using a spiral scaffold as a guide. It is these lines, which the spider generates from its flagelliform gland, that are more flexible, softer and are responsible for absorbing the impact when prey collides with the web. It also acts to ensure that the prey is caught and doesn't just simply buzz off after impact – even in wet weather. The radial lines are thicker and stronger than the connecting lines, and when the spider web is complete it is under tension with the radial thread under the maximum force. While radial silk is smooth and non-sticky, spiral silk threads are coated with a thin layer of glue. Thanks to the Rayleigh instability, which we came across in Chapter 2, this coating breaks into uniformly distributed droplets, forming a unique bead chain. By modelling the properties of the spider web, scientists have found that, in a typical example, a spider can change the number of radial or spiral threads without reducing the overall strength.[15] This versatility is important given the many different environments in which a spider must build its web-like home.

Being able to catch a fly is one thing, but knowing where it is on the web is another matter. For anyone who has seen a spider in action, the critters are remarkably quick at dealing with an insect once it has landed on the web. When an unfortunate victim gets caught, the web vibrates, with the ripples propagating through the structure. The transmitted frequency of a trapped insect varies between 50 and 100Hz due to the leg movements, and between 100Hz and 300Hz due to fluttering bees and flies. The frequency of such signals differs from those caused by the wind, which are typically below 10Hz. Spiders detect web vibration through their legs via sensory organs on their exoskeletons that detect leg displacement. There are usually two main methods to detect the vibration of a web. The European garden spider (*Araneus diadematus*) spends most of its time on the centre of the web, waiting, with all eight legs spread out so that it can detect prey wherever it is located on the web. The missing sector orb weaver (*Zygiella x-notata*), meanwhile, monitors its web from a distance and waits for vibration on a single thread that it places on one or two legs and is connected to the centre of the web. When an insect is caught, the spider then moves to the centre to orient itself and detect where the prey is located. In 2019, researchers in Spain and the UK used computer models of webs spun by orb-weaving spiders to investigate whether vibration alone could contain the necessary information for spiders to locate prey. They found that the spider can compare not only the magnitude of the vibration across its eight legs but also the distance and direction of any insect caught.[16] There really is no escape for that unfortunate fly.

Another common type of spider web is the tangle web. This appears to be a messy three-dimensional jumble of threads compared to the seemingly straightforward two-dimensional nature of the orb web. In 2021, researchers in the US and Germany put a tangle web spider (*Tidarren sisyphoides*) inside a transparent box and illuminated the box with a laser. They then watched day by day as the spider built its web, finding that the foundation of the web geometry is mostly complete after the first two days. For the next five days the spider then reinforces the existing network of fibres by increasing the web's density to boost its strength and toughness without increasing the overall size much further.[17] By throwing 'prey projectiles' at the webs, starting from day one, they found that, even by that time, if prey landed in the right (or should that be wrong?) spot then it would be caught in the web, even though it was not complete and had a fair number of gaping holes.

What all this work shows is that 100 million years of evolution must count for something, whether it is a flying insect or one waiting for the next meal. As the twentieth-century Soviet engineer and writer Genrich Altshuller once remarked: 'In nature there are lots of hidden patents.'

And finally, this book is about how you can use physics in your everyday life. So here is one piece of advice about tackling a major pest in the garden: snails.* For some reason, our garden is always inundated with these gastropods. Lift up a stone and there will be tens of the blighters underneath. The growth of snails could be getting worse thanks to climate change, with mild winters and warmer, wet springs being ideal for their reproduction and survival. So, according to physics, what is the best way to protect your garden from this climate-related disaster? Of course, killing the snails

---

* The UK's Royal Horticultural Society no longer refers to slugs and snails as pests, stating that they are a 'normal part of the garden ecosystem'.

outright is one option. But what if you don't want to subject your mollusc friends to the underside of your shoe and have already tried everything else, such as broken eggshells, sandpaper and copper wiring around the prized rose bush? In 2014, David Dunstan, from Queen Mary University of London, devised anther way – and one that is perhaps guaranteed to work.[18]

Dunstan's garden was looking 'shabby', so his friends helped him to choose about 100 new plants, which he then spent days carefully planting and nurturing. Soon after, however, and to his horror, the new garden was a wreck, with most plants, according to Dunstan, 'suffering severe snail damage'. He didn't want to spend all his time simply killing the snails and was told that if he threw them over his wall into a nearby wasteland, they would simply come back. But would they? He decided to investigate and went to some extreme lengths to find out, painstakingly marking 416 snails with Tippex spots. 'The snails had to be dry before applying the Tippex, so sometimes we had twenty to thirty snails on the table indoors,' Dunsten told me. 'I was astonished how quickly they could move.' The researchers then put the snails at various distances away from a central point to see how many and in what manner they returned. Using techniques from statistical physics, they found that the snails did not move in a random way but, given how many came back, seemed to 'drift' under the influence of a homing instinct. Looking deeper into the data, they found that the snails formed two groups: one that wasn't really bothered about returning and another that seemed hell bent on coming back to munch on the flowers.

So, how do you finally clear your garden of snails according to physics? The duo found that lobbing a snail about 5m away from your garden is enough to get rid of them for a month – but they will eventually make their way back. The truly magic distance, however, is 20m, from which point even the keenest very rarely returned. All you need to do to rid your garden of snails is to develop a strong throwing arm. Just make sure the neighbours don't catch you lobbing them on to their cherished vegetable patch.

6

PLANES, TRAINS AND AUTOMOBILES

A SIGNIFICANT PART OF the day involves getting out and about, whether going to school, commuting to work or visiting the local shops. While there are many ways to do this, such as walking, cycling, scooting or taking public transport, most of us (especially in the UK) take the car. The average person in the UK made 750 journeys in 2022, half of which were done by car or van (447 trips), with the remainder made up by walking (235), by bus (27), by train (19), by cycling (15) or by taxi (7).[1] While a decade ago this was terrible for the environment, it is not quite as bad today thanks to the emergence of electric vehicles, and the introduction of congestion charges and ultra-low emission zones, which are springing up in many cities across the UK to help drive down pollution.

Despite that, people today still struggle to decide whether to buy a petrol or electric vehicle (after all, modern petrol, and even diesel, cars are deemed 'clean' regarding emissions). Petrol cars are currently cheaper to buy upfront but come with more volatile running costs and are potentially more expensive to operate in the long run. Huge strides have been made in the past decade with electric cars, which now have lower battery costs together with travel ranges of more than 400km before the batteries need recharging. There is also an efficiency advantage, given that an electric motor converts about 70 per cent of the energy it consumes into mechanical energy, compared to the 30 per cent or so efficiency of a conventional internal combustion engine in a car. Yet electric cars might not be so environmentally friendly if the energy used to charge them doesn't come from green power sources such as renewables.

In 2023, physicist René Ledesma-Alonso, from the National Autonomous University of Mexico in Mexico City, and his colleague Guillermo Becerra-Nuñez undertook a cost–benefit analysis of buying a petrol versus an electric car. 'I was thinking about buying a new car, and as the popularity of hybrid and electric vehicles is growing, I wanted to analyse the advantages they have over gasoline vehicles,' Ledesma-Alonso told me. Using data relevant to Mexico City, they considered how often someone would travel, their average speed, time spent in traffic, vehicle taxes, as well as maintenance and service costs – considering both historical data as well as what is predicted in the future.[2] They found that whether to go for an electric or petrol car comes down to vehicle usage and the type of traffic conditions that are most frequently encountered. Electric was found to be best in heavy traffic and when weekly commuting distances were long (over 200km per week). Petrol, on the other hand, was found to be good for shorter distances and in light traffic. If you live in a traffic-prone area, such as a city, then electric could be the way to go – and the likely more environmentally friendly option, too.[*] Ledesma-Alonso adds that their model can easily be applied to other cities and says that before making any choice, think about your needs and budget. As for him, he ended up buying an electric car even though he calculates it would take six years for the running costs to be on par with a petrol vehicle. 'One advantage is that in Mexico City there are policies to restrict the use of gasoline vehicles when pollution levels are high,' says Ledesma-Alonso. 'With an electric vehicle, I don't have to worry about that.'

Another concern for electric car enthusiasts is what kind of price (or losses) you might rack up when reselling the car.[†] This usually depends on several factors, such as the vehicle make, model, milage and the condition of the car. For electric vehicles, resale value can also be affected by the driving

---

[*] There could be a dangerous downside to electric vehicles: their quietness. A study in 2024 of casualty rates in the UK between 2013 and 2017 found that as a pedestrian you are more than twice as likely to be hit by an electric/hybrid car as compared to a petrol/diesel vehicle; see Edwards, P.J., Moore, S. and Higgins, C., 'Pedestrian Safety on the Road to Net Zero: Cross-sectional Study of Collisions with Electric and Hybrid-Electric Cars in Great Britain', *Journal of Epidemiology & Community Health*, doi: 10.1136/jech-2024-221902.

[†] A study in 2024 of car drivers in the Netherlands found that there was more chance of having an accident when driving an electric vehicle compared with a conventional petrol or diesel car; see McDonnell, K., Sheehan, B., Murphy, F., et al., 'Are Electric Vehicles Riskier? A Comparative Study of Driving Behaviour and Insurance Claims for Internal Combustion Engine, Hybrid and Electric Vehicles', *Accident Analysis & Prevention*, vol. 207 (2024): 107761.

range and the battery condition. In 2023, researchers in the US examined more than 9 million car listings at over 60,000 US car dealerships between 2016 and 2022.[3] While they saw little difference in depreciation between conventional cars and hybrid vehicles, fully electric vehicles did not hold their value as compared to conventional cars. This is especially so for older models of electric vehicles that have shorter driving ranges. Tesla cars were an exception, however, holding their value much better when measured against other electric car manufacturers. The researchers found that things are beginning to change as electric cars with larger ranges are introduced. The latest Tesla models were found to follow the depreciation rate of conventional cars, which for anyone who has bought a new or nearly new petrol or diesel car already knows is bad enough.

When it comes to travelling, car is still king, despite the environmental and health benefits that other options may bring. Yet each day millions of drivers sit motionless in traffic either on their way to work* or on busy motorways trying to get away for the weekend. No one likes traffic but it is a fact of life for many and there is no worse feeling than driving freely on a motorway one minute only to come to a grinding halt the next. It doesn't take Albert Einstein to work out that traffic jams are, in a simple sense, caused by too many vehicles being on a certain stretch of road at a given time, or a high 'vehicle density'. Surely, no rules or laws can be garnered from all this? Yet for the past

---

* The average commuting time in the UK is twenty-seven minutes, with estimates that people waste more than sixty hours a year sat in traffic. Grim.

seventy years scientists have tried to understand the mathematical rules of traffic flow, modelling it as individual non-interacting (hopefully!) particles moving in one direction along a 'road'. If the vehicle density moving along a motorway with multiple lanes is low, then cars can easily overtake each other, resulting in 'free flow'. As you start to add more cars, free-flow traffic can move into another traffic state called 'synchronised flow'. This is where traffic is heavy but still flowing at a similar speed. When the vehicle density reaches a critical limit, in other words free-flowing or synchronised-flow traffic comes up to a bottleneck caused by an accident or a broken-down vehicle, then cars start to slow down, and the result is a traffic jam that moves further 'upstream'* over time as more cars join the queue.

Traffic jams on the motorway are usually due to accidents, roadworks, a broken-down vehicle, or from general congestion at a particular pinch-point like a junction. Yet there is also evidence it is a bit more nuanced, with drivers causing traffic jams to spontaneously form (without crashing). In 1997, researchers in Germany studied traffic on a busy section of the A5 southbound motorway near Frankfurt, Germany. This road, which is one of the busiest in Germany, connects Giessen in central Germany with Basel in Switzerland, and gets particularly busy near Frankfurt. The researchers put 'induction-loop' detectors on the road at different points to measure the speed of every vehicle as well as to determine the flow of cars through those different points. While

---

* It might be slightly confusing, but the back of the jam where cars enter is termed 'upstream', while the place where cars exit the jam is termed 'downstream'.

they found that a spike in the density of vehicles at a specific point like joining a motorway can lead to a jam further upstream,[4] they also discovered that 'small perturbations', or tiny changes in vehicle motion, could equally lead to jams.

From the data alone, the researchers couldn't identify what led to these 'perturbations' but suspected it was caused by a single driver getting distracted, such as needing to hit the brakes to avoid crashing into another car or veering slightly off the road and having to correct themselves. In dense traffic, if a driver gets distracted for a split second and comes too near the car in front, for example, they then need to hit the brakes. This might mean the car behind has to do the same and brake harder to stop careering into the car in front. If this action is replicated among all or most drivers on a road or motorway it results in a so-called 'phantom' jam. It is named so because there is no obvious reason for the jam to occur other than in the behaviour of a single or a few drivers and thus can't be predicted purely from the number of vehicles alone, although it is much more likely to happen if the distance between cars is less than 35m. These phantom disruptions move like a wave through traffic and can travel for miles. And given that all physicists love waves, they have deemed the movements 'jamitons'.[5] As the waves travel slower than the moving cars, it means that those cars upstream of the wave always catch up and thus must slow down or even stop in the traffic jam before accelerating off again. This is that annoying stop–start style of jam, in which all vehicles are happily going along, only to suddenly come to a stop, before being freed but then hitting (not literally) another jam.

An experiment carried out in Japan in 2008 showed that the phantom traffic jam could even occur in an incredibly simple setting. Twenty-two cars were placed equally spaced apart on a 230m-circumference circular track. Each driver was given a simple instruction: maintain a constant speed of 30km/h and a fixed distance to the car in front.* Easy enough you might think, but apparently not. Initially, the cars were all moving along smoothly and roughly keeping the space between them the same, but after a while, cars began to form clumps. Traffic jams then ensued, reaching such an extent that even in this simple road layout with slow-moving cars, a clear stop-and-go wave formed. It also exactly matched the theoretical prediction of a jamiton's behaviour.[6] The wave moved at a rate of about 20km/h – a common value seen in real-life motorway traffic jams.

Given that humans were responsible for causing these jams to occur – perhaps a nervous driver not paying attention or someone being a bit bored and careless – this opens the possibility that in future self-driving cars and the complex algorithms that operate them may offer a solution. They may be able to control both speed and distance relative to other cars without momentarily having a lapse in concentration (one would hope). They could also consider what is happening kilometres in front of them via real-time traffic data to maintain a certain speed to keep traffic moving. One recent move in the UK to tackle the impact of traffic jams and congestion is via variable speed limits on motorways. If there is a lane closure caused by roadworks or an accident, for example, it means that vehicles in three lanes must merge into two. Vehicles on the motorway that are making their way towards the accident/issue are told that there is a lane closure and are made to slow down

---

* For a video of the jam in action, see youtu.be/Suugn-p5C1M.

and keep a constant speed. This, it is thought, helps to increase road capacity and prevents stop–start waves from occurring ahead. Yet, self-driving cars may do this automatically while also communicating with other cars to optimise speed, route and distance between them. And it might not take that many to solve the traffic-jam problem. An experiment in 2018, again using a circular road with about twenty cars on it,[*] included a twist — it had a single 'control vehicle' that was used to manage the traffic. Even in this case it was possible to remove phantom jams and the traffic waves they produce.[7] Extrapolating the results means that if only 5 per cent of vehicles have some sort of artificial intelligence traffic-management capabilities[†] then it could be enough to remove phantom jams entirely. Indeed, a follow-up demonstration in 2022 with 100 semi-autonomous vehicles on a real highway outside Nashville, Tennessee, showed that the real-world deployment of vehicles with AI traffic management is feasible and safe, and that existing technologies, such as adaptive cruise-control systems, can be useful to tackle phantom jams.[‡]

Perhaps 'intelligent' self-driving cars may finally allow us to say 'goodbye gridlock'.

---

[*] For a video of the experiment, see youtu.be/2mBjYZTeaTc.

[†] It has been shown that typical adaptive cruise-control systems alone are not sufficient to control traffic flows; see Gunter, G., Gloudemans, D., Stern, R.E., et al., 'Are Commercially Implemented Adaptive Cruise Control Systems String Stable?', *IEEE Transactions on Intelligent Transportation Systems*, vol. 22 (2021): 6992.

[‡] See youtu.be/PA3lyoCZnP0.

Destination reached, it's time to park the car. It's a classic conundrum for drivers looking to park at a popular or busy destination, say a football match or concert. Should a driver park the vehicle far from the destination, where finding a spot may be easier but has the downside of the need to walk a longer distance, or instead risk trying to park as close to the destination as possible, where spaces are harder to find but with a chance of just getting lucky? The approach you take might come down to your personality: are you a risk taker or do you like to take the easy, trouble-free route (like I do). This taxing issue was tackled in 2019 by physicists Paul Krapivsky of Boston University and Sid Redner of the Santa Fe Institute. They applied methods from statistical physics to the problem of a single street leading up to a destination.[8] They looked at three strategies under the assumption that a driver cannot see empty spots ahead, only what is directly in front of them. If a driver approaches a three-space gap, for example, they can see all three spaces but nothing further. While this might be totally unrealistic, let's pretend it is a very foggy day. 'There's a famous quote that "models should be as simple as possible, and no simpler",' Redner told me. 'That's what motivated me to boil this down to a one-dimensional model.' A 'meek' strategy involves parking at the very first available spot, which wastes no time when parking, but also happens to be inefficient if there are spaces available nearer to the destination. Nobody usually does this because most people will hope that a better space is available. A 'prudent' strategy involves passing that first spot (or group of parked cars) and then parking the car in the next best spot available (i.e., if after that first free spot there is a parked car and then a group of three free spots, parking in the one nearest to the venue). The potential issue with the prudent strategy is that if there isn't another space after that first free spot of the meek strategy, you end up going all the way to the end of the road looking for a space before eventually backtracking all the way back to the meek position, thus wasting time. An 'optimistic' strategy, meanwhile, involves driving all the way to the target location, ignoring any empty spots on the way, and then backtracking away from the destination and parking in the first available spot. Now you might think that the optimistic strategy is best, as it always guarantees the closest spot, but when the researchers ran simulations on all three scenarios, they found that in this admittedly rather simplistic model, the prudent strategy was the best way to save time. The downside of the optimistic strategy being that you risk a lot of time backtracking, not to mention the stress of being late for kick-off or that dinner date.

In reality, many of us instead would look as far as we can into the distance and park in the furthest free spot we can see. Then, once we arrive, if there is a further free place available nearer to the destination, drive to it instead. Repeat, otherwise park. Yet, a year later, Redner and Krapivsky were back for more parking theory and attempted to optimise the solution. Again, they considered a single street parking lot. In this case, however, they wanted to know what was the optimal fraction of the total length of road to bypass, even though there were spaces available, to give you the best chance of landing the perfect parking spot? While for 'optimistic' drivers this fraction would be one and for the 'meek' it would be zero, one might think that the truly optimal strategy would be somewhere in between.* And this is exactly what they found – you increase your chances of finding the best parking spot by ignoring at least half the spaces (occupied or unoccupied) before parking in the next available space.† If you ignore more than half, then you risk not finding a space, and ignoring much less than half risks parking too far away from better spots ahead. By carrying out this procedure, the probability of finding *the* best parking space is 25 per cent.[9]

Will this finding persuade you to change your parking strategy next time you visit the shops? Probably not. 'With regard to parking, humans do not follow optimal strategies,' the researchers state. Yet Redner says that because of his work he finds that he is now more 'cognitive of the optimal strategy' and that it can pay to ignore some free spots for a better one ahead. 'The only thing is, I almost never drive a car,' he adds.

While cars are used for more than 80 per cent of journeys in most towns and cities across the UK, in London it is only 27 per cent, mostly because of the London Underground (and bus network). I did my PhD in Stuttgart, Germany – a country that is renowned and envied for its public transport system. Stuttgart has both overground and underground

---

* This 'threshold' strategy is linked with the classic 'secretary problem', which we will encounter in Chapter 13.

† A slight caveat being that this solution only holds when all drivers use this strategy.

trains (S- and U-Bahn) connecting most parts of the city, as well as a bus and tram network. They all ran like clockwork – any delay, even to national train journeys, was a surprise and likely the result of something significant happening. One aspect I noticed when I lived in Germany compared to the UK was the timeliness of the bus network. This was likely because of alternative methods of transport, such as tram or S- and U-Bahn, which helped to alleviate demand for buses as well as take cars off the roads, particularly during busy times.*

In the UK, all those cars during rush hour means it is difficult for buses to maintain their schedules during weekday mornings and evenings. This is also likely behind the saying 'you wait ages for one bus and then three come along at once'. And it is not just a myth but really does happen. The explanation for why it occurs is straightforward enough. Let's say that each bus on a particular line starts off equally spaced. If there is little traffic or demand, then they should be able to maintain their time gap and schedule (such as during the day). But rush-hour traffic conditions or even perhaps a spike in the number of passengers at a particular stop (such as outside a train station near a football stadium after a match) can change things. This spike in passenger numbers at busy times means that the first bus needs more time to pick up passengers. It gets delayed and the bus behind starts to catch up. The delay for the first bus means that more passengers arrive at the bus's next stop, which

---

* What was once a source of national pride, in recent years the German national rail system has been hit by frequent delays and cancellations; see www.nytimes.com/2024/06/23/world/europe/germany-trains-euro-2024.html.

delays the bus even more and the bus behind now catches up as it has fewer, if any, passengers to pick up. This creates a 'platooning' of two buses. But it also means that the third bus catches up too, given the lack of passengers caused by the first and second bus. When simulations are run on such a network it results in three buses platooning together and then a large gap between this platoon and a fourth bus (as this fourth bus is picking up passengers that bus two or three should have done if they weren't so speedy). Buses really do come in threes when there is a delay, such as a spike in passengers or heavy traffic conditions. But it can have some advantages. If a bus is delayed and it is full there is a good chance that an empty one is close behind, so it can pay off to wait for it. This doesn't just help you to get on an empty bus so you can sit in your favourite seat, but it can go some way to helping the whole system get back on track, as it aids the bus in front to get ahead of the forming platoon.

For public transport operators, one option to get around the bunching problem is to let the bus behind overtake, but this doesn't solve the issue: they just end up leapfrogging each other at every stop. This is also not possible for metro systems that can also be impacted by bunching. Timetable operators may instead put more capacity at peak times, or to make up time, as they do for some metro systems, a delayed train will announce that it will not stop at one or two stations ahead. But is there a better way to avoid wasting infrastructure and fuel, not to mention annoying passengers? In 2009, systems complexity researchers Carlos Gershenson and Luis Pineda at the National Autonomous University of Mexico wanted to find out. They created a network model that included five stations, five vehicles and a track length of a 120 vehicles. Passenger numbers at each stop were random. They showed that allowing empty trains to overtake didn't solve the issue, while adding more trains during heavy periods led to longer platoons.[10] The only way to tackle the problem was to not allow any gap to develop. To do this, trains spent a minimum amount of time at each station, which meant that they didn't speed up and catch up to the train in front. They avoided being delayed by leaving the station after a certain amount of time, even if boarding was not fully complete. If these conditions were met, then platoons never formed.

Two years later, Gershenson came up with a modification that solved the bunching problem directly.[11] In this case, the model was based on 'self-organisation' and considered the live status of passenger numbers at stations. It then used that information, together with details about when the train in front departed the station and the distance to the train behind it, to determine when to leave the station. If, for example, the train in front left the station a long time ago, then the current train at the station doesn't spend long picking up passengers to avoid too large a gap from occurring. That's the theory, but is it possible in practice? In 2017, the researchers tested their model using real data from Mexico City Underground, which serves some 6 million users every day on twelve lines. 'It was tough to get agreement to do this,' Gershenson told me. 'It took two years and was a competition between the funding agency, the university and the metro authorities to see who had more bureaucracy.' Once they got the green light, they began taking passenger and train data at Balderas station, which is a transfer station in central Mexico City along lines 1 and 3. This included information about the number of passengers, train arrival and departure times, and whether there was any delay. They found that a significant impact on train timeliness came from overcrowding as well as from passenger behaviour, such as trying to get on a train that was already full or stopping the doors from closing, which in each instance caused a delay of about a minute.[12]

They then created a model that consisted of a railway line of almost 20km, which serviced twenty stations each with a length of a 150m and with sixteen trains on the track. By maintaining a minimum and maximum waiting time for trains at the stations in their model, they were able to decrease journey time by some 20 per cent and the time between trains by a quarter over what they saw happen at Balderas station. Those benefits rely on passengers playing along and not acting in a way that delays trains. So the researchers came up with a set of recommendations for 'good' metro behaviour, such as being patient when boarding and making sure you let people get off easily by making space. Doing otherwise means that it could delay everyone, impacting the waiting time at a stop and increasing the chance of bunching behaviour. The researchers implemented some of these suggestions at Balderas station by putting arrows for passengers on the floor to guide those who were disembarking from a train and those waiting to get on – what you might already see in other metro stations around the world, including Tokyo, Beijing, Munich and Dubai. They found that this reduced boarding and alighting time by about 10–15 per cent, with cases of door obstructions being reduced by around 20 per cent. This was so successful that it was then rolled out to other stations in the network, which saw immediate benefits.

Another tip the researchers have is that on a crowded vehicle go as far away from the doors as you can, as this will allow space for people to embark and disembark from the bus or train, speeding up the process. 'There are many small things that can be done by an individual to improve the overall flow,' says Gershenson. Like car drivers, individual behaviour really can matter, with 'bad' actions having a 'butterfly' effect on the whole system, making it come to a screeching halt.

As well as having a good public transport network, many European cities, such as Utrecht in the Netherlands or Copenhagen in Denmark, are cycling hubs. Rental electric bikes or scooters have boomed across major European cities in recent years, following a similar trend in Asia. In many cities it is hard to avoid the brightly coloured electric bikes or scooters strewn across pavements. Electric bikes and scooters not only provide a solution for traffic and parking problems but also for the environment.* A 2018 study of rental-bike usage in Shanghai estimated that their introduction saved some 8,000 tonnes of petrol in 2016 as well as reduced carbon dioxide emissions by 25,000 tonnes

---

* While getting on your bike or scooter is a good thing for the environment, there are potential downsides: accidents. In the US there were about 23,500 e-bicycle injuries in 2022, up from just 751 in 2017 while there were about 8,500 e-scooter injuries in 2017 but almost 57,000 in 2022; see Fernandez, A.N., Li, K.D., Patel, H.V. et al., 'Injuries with Electric vs Conventional Scooters and Bicycles', *JAMA Network Open*, vol. 7, no. 7 (2024): e2424131.

and limited the release of nitrogen oxides by more than 60 tonnes.[13]

Users rent the scooter or bike with their mobile phones and once their journey is complete, dock it and pay for the usage time. A typical charge may be £1 to unlock the bike and then 20p per minute to use. An analysis of e-bike users in Dublin, Ireland, in 2022 showed that users rent a bike for about fifteen minutes, on average, travelling some 250m, with some users going as far as several kilometres (did they get lost?). But the schemes are also open to abuse. About 13 per cent of users did not park their bikes correctly in the designated parking areas following use.[14] This was a particular problem in China, where some companies did not stipulate where users should park their bikes. This led to people abandoning the bikes everywhere, resulting in 'bike graveyards' at popular destinations.*

Indeed, a problem faced by rental e-bike and e-scooter firms is rebalancing the location of bikes or scooters. After all, demand is high in the morning to get into the city centre, for example, while in the evening most people will rent one to go home after work, with another contingent needing a bike or scooter to travel into the centre in the evening for dinner or drinks. In 2022, scientists at Tokyo University of Science looked at bicycle usage in four US cities: Boston, Washington DC, New York City and Chicago. These cities have more than 560 ports each (except for Boston, which has 320) for a total number of bicycles in the thousands. They found that most usage, as expected, happens during the day

---

* See www.theguardian.com/world/2017/jan/17/chinese-discard-hundreds-of-cycles-for-hire-in-giant-pile.

and that usage patterns were similar between the weekday and weekend.[15] They also discovered both an excess and a lack of bikes across these four cities all year round. Some ports eventually became empty and others became full, meaning that people either couldn't rent a bike or return one, leading to disgruntled users. Many firms use a fleet of trucks to reallocate dozens of bikes at a time, but this can be costly and time-consuming if done in a haphazard manner. In some cases, drivers just travel around cities looking to pick up or drop off bikes, doing so on feeling or experience.

This rebalancing act is different depending on the city. Some cities, such as New York, tend to require much less balancing compared to cities that are hillier (in other words, people like to rent a bike to go up the hill but might instead walk back down). While algorithms and analytics are being developed to tackle this challenge, it is far from straightforward, as the problem is dynamic, with the rental 'landscape' of bikes/scooters always changing. It is also random: it can be difficult to predict when a customer will pick up a ride and where they might go. But that hasn't stopped scientists from trying. In 2022, Steffen Bakker and colleagues at the Norwegian University of Science and Technology in Trondheim mined usage data of bikes and scooters in Oslo to develop a model that considers historical data as well as what has happened so far during the day to predict what will happen in the next few hours. The algorithm calculates a 'criticality score' for each station, or a measure of expected demand in the near future, and provides recommendations about what service operators should do – whether they should drop off or pick up bikes at a specific station, swap batteries, and where to go next.[16] They calculate that the algorithm could reduce problems such as empty or full stations by up to 40 per cent compared to not doing anything. When they then compared it to the practices of a firm in Oslo during 2017 – at the time there were 158 stations with 1,790 bikes in use – the program could reduce the number of 'service issues' by up to a quarter. Buoyed by the findings, the team plan to test the algorithm directly in Oslo via an app given to drivers. 'I am confident we can improve the way bikes are managed,' Bakker told me, who is a keen cyclist but uses his own bike.

If you find it easy to get a ride with a scooter or bike next time, it might be thanks to some clever optimisation algorithms.

Flying is admittedly not an everyday occurrence, although some people are frequent flyers for work or perhaps pleasure. The annoying aspect about going to the airport is all the waiting – waiting to check in, waiting for security check and passport control, waiting for the flight to arrive, waiting to board, waiting to take off, waiting to get off the plane, waiting for passport control, waiting for baggage – you get the drift. When it comes to waiting to board, many airlines board passengers back to front, calling certain groups of seat numbers that start with a bunch at the back. This might make sense given that those people that have boarded can't then get in the way of the next group. You might not think there is a lot of physics in all this waiting, but there are ways that physics can help to alleviate it. In 2005, astrophysicist Jason Steffen, who at the time was an astrophysics graduate student at the University of Washington, was travelling from Seattle airport when he became frustrated by the long wait to board the aeroplane as everyone stood around, either stowing their luggage or waiting for others to do so.

Steffen began to think how boarding an aircraft could be done better. It consumed so much of his time that he gave himself an ultimatum: solve the problem or move on with life. He decided on the former and then devised an aeroplane-boarding 'theory' that consisted of an optimisation algorithm* (like the e-bike problem). He then ran subsequent simulations using a mock aircraft, which contained twenty rows of six seats with one aisle running up the middle. There was no first-class cabin, no priority seating and each flight was full. He also presumed that stowing away luggage, which he dubs as being 'aisle interference', takes up the bulk of passenger boarding time, while other issues such as passing someone who is already seated does not cause any delay.

Steffen found that the worst way to board an aeroplane is front to back, which is to be expected, but he also found that the current airline standard – back to front – is the second worst way to board and is almost as bad as front to back.[17] Steffen discovered that the best way for passengers to board is back to front, but in such a way that adjacent passengers in the queue are seated two rows apart (12A followed by 10A and then 8A, for example). In this scenario, passengers would first fill out all the right-hand side of the aircraft. Those that have window seats with even numbers would be seated first (from back to front of the aeroplane), then middle seats, then aisle seats. Next, the same would happen

---

* The algorithm is based on a Markov chain Monte Carlo simulation.

for odd row numbers – first window, then middle, then aisle. This process would then repeat for the left-hand side of the aircraft. In this case, the simulation showed that the boarding time would be four to ten times faster than boarding in blocks from front to back. 'The reason my method is fastest is that it optimally uses the aisle,' Steffen, who is now at the University of Nevada, Las Vegas, told me. 'People have sufficient personal space between themselves and adjacent passengers, but otherwise the aisle is filled to capacity with people who can put their luggage away and sit down.'

Of course, this works well if everyone is an individual traveller, but that is not always the case, as there are usually families who must board and sit together (even if they would like to be sat as far away as possible to have a few hours of quiet). So, Steffen proposed a modified version, in which passengers board in blocks of three consecutive seats on one side of the plane in every other row, which would result in four boarding groups. This proved a good compromise: it is twice as slow as the optimal method but twice as fast as the conventional method. Interestingly, some airlines employ 'free-for-all' boarding, where passengers can sit anywhere. You might think that this would be chaotic, but in another study in 2008 Steffen applied techniques from statistical mechanics to a random boarding strategy and found that it is almost as good as the optimal boarding method.[18] It also has the advantage of not needing the airline to organise passengers.

When Steffen published the optimal boarding strategy it caught the attention of writer, producer and director Jon Hotchkiss of Hotchkiss Industries in California. The pair then worked together on testing the theory in real life.

To do so they used a mock fuselage of a Boeing 757 that was located at the Air Hollywood soundstage in Studio City, California. The aircraft had twelve rows of six seats with a central aisle. They employed seventy-two 'passengers' aged between 5 to 65. Passengers either had no luggage, a bag or a cabin-sized suitcase, or in some cases both. The amount of luggage was such that when all the passengers were aboard, the overhead bins were full.*

Each passenger was then given a ticket with a seat assigned and either a passenger number or boarding group number, but they did not know which boarding method was being tested. The pair then tested five strategies: Steffen's optimal method; back to front; four-row blocks; random boarding; and 'Wilma', in which windows board first, then middle seats, then aisle. To make it more realistic, they boarded passengers who had children first. They found that the four-row blocks method was the worst to complete the boarding, which took six minutes and fifty-four seconds, followed by back to front, which took six minutes and eleven seconds. Next came random at four minutes and forty-four seconds, followed by Wilma at four minutes and thirteen seconds. Steffen's optimal method only took three and a half minutes.[19]

Steffen says that the bigger aircraft, the bigger the time savings, so if the aeroplane is twice as long, the time savings are almost twice as much. For the airline industry time is, after all, money and Steffen calculates that the savings for an airline using his optimal method over the block method could be about $1.1 billion per year. That could be why

---

* To watch a video of the boarding, see youtu.be/u85SzI6KsBQ.

some airlines are now rethinking the status quo of boarding from the back. In October 2023, United Airlines announced that it would start boarding economy-class passengers on domestic flights using the Wilma method.[20] The airline says that first- and business-class passengers would see no difference, but the change would save up to two minutes of boarding time per flight over its usual practices.

And when it comes to waiting at the airport with incredibly impatient children, every minute counts.

LIFE'S A SPORT

PLAYING A SPORT OR doing some exercise is an important part of the day for most people. Exercise, we are told, is good for you, not only from a physical health perspective, but also for your mental health by reducing stress and boosting self-confidence. Some may prefer to get their exercise over with as soon as possible and head to the gym in the morning for a quick cardio or weights session or go for a jog/swim to sharpen the mind for the day ahead. Others, meanwhile, might prefer a lunchtime walk to reduce stress and help them focus on the afternoon. And some, like me, may prefer instead to leave exercise to the evening and head out to the local tennis court or football pitch to combine exercise with socialising.

Participating in a team sport such as cricket or rugby helps you to run off some calories as well as to make friends and improve your teamwork and discipline skills. And when it comes to the most popular sport worldwide, it's hard to beat football. With an estimated fanbase of 3.5 billion, it is big business, with top football players on contracts worth millions of pounds each year, which make them some of the best-paid athletes on the planet. Between 2019 and 2022, football's governing body, FIFA, had revenues of $7.6 billion, most of which came thanks to selling sponsorship and lucrative broadcast rights for the 2022 World Cup in Qatar.[1] The top twenty clubs in Europe, meanwhile, made €11.2 billion in the 2023/24 season – an increase of 6 per cent on the previous year. This was mostly thanks to record attendances at games, with matchday revenue reaching €2.1 billion. In 2024, five of the English Premier League's 'big six' clubs,* saw an increase of 15 per cent or more in revenues thanks to new commercial partnerships and an increase in non-matchday events, such as concerts and stadium tours.[2]

All that money sloshing around football is partly because it is part of the lives of so many people who spend their hard-earned cash on merchandise and season tickets. The sport is woven into the very fabric of towns and cities across the UK (and across many other countries), with stadia representing the cathedrals of our time, which each week welcome tens of thousands of fans to support their team. One of the most famous stadia in the world is Camp Nou, home to FC Barcelona. Having opened in 1957, in the early 1980s it had an epic maximum capacity of 115,000. Even after standing areas were replaced with seating, that figure today is 99,354, with every home game a sell-out.

If you live near a stadium and a goal is scored by the home team, you will likely hear the roar of thousands

---

* That is Arsenal, Chelsea, Liverpool, Manchester City, Manchester United and Tottenham Hotspur.

of fans celebrating. Large stadia, including Camp Nou, not only have an acoustic impact but a seismic one, too. In 2016, seismologist Jordi Díaz from the Institute of Earth Sciences Jaume Almera in Barcelona (now known as Geosciences Barcelona) installed a seismometer in the basement of the institute's building, which lies a few hundred metres away from Camp Nou. As soon as he turned on the device, Díaz could pick up signals from football fans jumping up and down when their team scored, and could even tell from looking at the seismographs when the almost 100,000 fans entered and left the ground. Yet an untypically large signal occurred on 8 March 2017 when FC Barcelona hosted Paris Saint-Germain (PSG) in the second leg of a UEFA Champions League knock-out game. Trailing four–nil from the first leg in Paris, Barcelona needed to at least match the score to take the game into extra time. They began well, going three–nil up inside fifteen minutes, but disaster struck for Barcelona when PSG striker Edison Cavani scored in the sixty-second minute. The 'away goals' rule meant that Barcelona needed three more goals to win. Time was almost up as the game entered the ninetieth minute, but Barcelona's Neymar scored two goals in quick succession, taking it to five all on aggregate. With PSG still going through, Barcelona midfielder Sergi Roberto popped up in the final moments of the match and scored, taking Barcelona to the next round and sparking wild celebrations among the home fans. As a result of that goal, Díaz measured the biggest spike in the seismograph that he had ever measured during a game – a recording of 1 on the Richter scale. In effect, the supporters created a micro-earthquake in the Catalan city.[*] Even the other goals showed a clear signal spike above background levels – except, of course, when PSG scored.[†]

When Barcelona, or any other team with a large stadium and fanbase, score, then the whole city might shake, but seismic waves aren't the only type of oscillation that can be seen at football games and many other sports. The Mexican wave, which is also known as *La Ola*, is thought to have originated during the 1986 FIFA World Cup in Mexico. Spectators in one section rise from their seats, fling their arms up in the air and then sit back down again as the next section rise to repeat the action, with the 'standing up' wave moving around the stadium. In 2002 researchers in Hungary analysed video recordings of fourteen Mexican waves in different football stadia, which had a capacity of at least 50,000 people. They found that, for some reason, Mexican waves usually move clockwise around stadia (from the perspective of the spectator the wave is passing by) and typically move at a speed of 12m/s, or about twenty seats per second with a width of about fifteen seats.[3] They also discovered that initiating the

---

[*] It is unlikely that anyone would have felt the tremor, as such quakes can generally only be noticed above a magnitude of 3.0 on the Richer scale.

[†] And it is not just football games that can have a seismic effect. A 2023 Taylor Swift concert in Seattle, as part of her Eras tour, generated activity equivalent to a 2.3 magnitude earthquake as 144,000 fans 'shook it off'. Scientists in Los Angeles who studied one of her concerts in the city that year found that they could identify when forty-three of the forty-five songs were played just by looking at the recorded spectrograms. See Tepp, G., Stubailo, I., Kohler, M., et al., 'Shake to the Beat: Exploring the Seismic Signals and Stadium Response of Concerts and Music Fans', *Seismological Research Letters* doi:10.1785/0220230385.

waves only requires a few dozen people, the caveat being that the initiators have to be in different rows rather than sat along the same one. So, the next time you visit your favourite team and want to strike up a wave, get a few dozen other people on board who are seated in different rows. And perhaps also strike when the mood of the crowd might be willing, such as during a quiet moment in the game. Don't try it when there is a penalty, and also don't expect the away fans to get involved, especially if they are losing.

Football is played by 250 million people in more than 200 countries. Part of its popularity is due to the low barrier to entry – as the saying goes, you only need a ball and jumpers for goalposts. Other sports such as cricket and golf need more resources in terms of both equipment and space. Yet the great thing about sport is that there are so many different ones to choose from, catering to a wide range of capabilities and interests. This includes differing ball sizes, playing areas, player numbers and ball speeds. Sports that involve throwing, such as basketball, for example, mostly feature ball speeds of about 15m/s, while a football that is kicked hard can reach 60m/s. The addition of a racquet helps to produce quicker speeds, with a tennis ball moving at around 70m/s, while a golf club can yield a ball speed of 90m/s. The quickest, however, might come as a surprise: badminton. The current 'ball' speed record was set in April 2023 by India's Satwiksairaj Rankireddy at 157m/s – more than twice that of tennis.* The reason being that while

---

* For this purpose, a shuttlecock is defined as being a ball.

a tennis racquet is rigid, the badminton version is slender and can be bent more easily during a fast movement, with the elastic energy converting into kinetic energy during the stroke. This allows the badminton racquet to hit the shuttlecock at twice the speed than if struck by a rigid tennis racquet.

The size of sport fields or areas of play is usually thought to be defined by coincidence, playing the sport within the rules and then setting the pitch size, but it seems it is often linked in some way with the maximum speed of the ball. In 2014, David Quéré and colleagues from École Polytechnique in Palaiseau in the southern suburbs of Paris wanted to take a quantitative approach to why a football pitch is some 100m long or why a tennis court happens to be roughly 23 × 11m.[4] They examined thirteen sports: badminton, basketball, beach volleyball, football/soccer, golf, handball, hockey, rugby, table tennis, tennis, volleyball, water polo and ice hockey. The researchers calculated the maximum distance a 'ball' can travel by using properties such as its size, mass and maximum speed, and then compared this to the size of the pitch or field. The interesting thing is that there is a limit to the distance a ball can travel, which has very little to do with its initial speed when struck. Instead, it depends on the density of the air (which is roughly constant) and the properties of the ball, such as its size and mass. Quéré calls this an 'aerodynamic wall'. In other words, you can hit a shuttlecock as hard as you like, but it won't travel much further than 14m. Similarly, a football won't travel in the air more than 100m or a golf ball more than 200m.[*]

---

[*] Many professional golf players will be able to hit the ball further than that distance due to bounces and roll.

The team found, rather unsurprisingly, that the longer a ball's range, the larger the field. After all, it is no use playing football or golf on a pitch the size of a tennis court unless you can rein yourself in from hitting it as hard as possible. The opposite is also true that if you played football on a pitch ten times the ball's maximum range, it would be a boring game. 'In some sports there is a connection between the size of the field and the size of the aerodynamic wall,' Quéré told me. 'We think this is not a coincidence and it is interesting to think of sports as having this principle.' There are, however, a couple of exceptions. While squash has a much smaller field than predicted for the parameters of the ball, it is played with walls that keep the ball in play (the same can be said for jai alai or indoor football).

Quéré and colleagues next examined how much time the ball needs to move across the whole playing field if it is travelling at its maximum speed and compared it to a player's reaction time, which is typically around a second. If the time to move across the pitch is shorter than the reaction time, then the sport mostly involves quick reflexes and decisions. They dubbed these 'precision and reflex' sports, which means that precision is important to keep the ball on the field of play, while reflexes are essential to keep in the point. Examples of these include squash, jai alai, tennis, beach volleyball and table tennis. Sports such as football, softball, handball, lacrosse, golf and basketball, on the other hand, are the opposite. It takes much longer for the ball in these sports to move across the field of play at maximum speed compared to the reflexes of a particular person. Depending on the sport, it could take several shots to move across the whole field – a minimum of two or three in football, for example, and if you are really bad at golf, then tens of shots per hole. This usually means that the ability to pass the ball becomes important (if it is a co-operative sport). The researchers dubbed these 'target' sports, in which strategy, skill and sometimes teamwork are used to reach a target – the hole in golf, a goal in football or a basket in basketball.

It turns out that different sports vary across a spectrum of being target or precision and reflex. Football and basketball are firmly target sports, while baseball is still target but also has elements of precision given that you can hit a baseball out of the court in one shot. Likewise, in precision and reflex, squash and jai alai are firmly in this category, while beach volleyball is still in the category but also has target elements. Yet there is one sport that deviates from these two definitions: badminton. In this case it is generally easier to keep the shuttlecock within the limits of the field of play (thanks to a particularly efficient aerodynamic wall), so precision is not as critical, but reflexes still are. This makes badminton purely a reflex sport. There are also likely to be sports that are mostly precision, snooker being one possible example. If you think snooker is a sport, that is.

Whether reflex, precision or target, the common denominator that makes many sports fun to play is the use of spin to either keep the ball within the limits of the court (as in tennis) or to bamboozle the opponent with the movement of the ball in the air (such as tennis, but also football, baseball, table tennis or cricket). And when it comes to spin, physics is the name of the game. Indeed, many scientific giants have taken an interest in sports, or rather were intrigued by the interesting physics that can occur in ball games. Isaac Newton being one. He is famed for his work on gravity, which, as the legend goes, he discovered when an apple hit him on the head as he sat underneath a tree at his home at Woolsthorpe Manor. In 1672, at just 29 years old, Newton noted how a spinning ball happened to swerve in flight. He postulated that the reason is that the spin creates an unequal air pressure at either side of the ball as it moves through the air. He wrote about his findings:

I had often seen a Tennis ball, struck with an oblique Racket, describe such a curve line. For, a circular as well as a progressive motion being communicated to it by that stroak, its parts on that side, where the motions conspire, must press and beat the contiguous Air more violently than on the other, and there excite a reluctancy and reaction of the Air proportionably greater.[5]

Newton's observations were incredibly astute, and the idea that a pressure difference could be behind the effect turned out to be a good hunch. Yet it still took 200 more years before it was fully explained. The first inroads came in 1738 when the Dutch-born mathematician Daniel Bernoulli published a theory of how fluids behave when they are moving (air being considered a fluid). As we found out in the previous chapter, the principle helps to explain how an aeroplane keeps aloft by considering how the air travels over the object (in that case the wings) to provide a pressure difference and a lifting force. Building on this work, in the mid-nineteenth century, the German physicist Gustav Magnus investigated why spinning bullets deflect to one side, while non-spinning projectiles remain straight. He published his findings in an 1852 work entitled *Über die Abweichung der Geschosse* (*On the Deviation of Projectiles*) in which he discovered a lifting force, with its strength depending on the speed of the rotation. The explanation for this effect is what is now known as the Magnus force. When a ball moves in the air it is subjected to a drag force from air resistance in the opposite direction to the direction of the ball (see figure below). If a ball is spinning on an axis, air travels faster along the surface of the ball that is spinning in the same direction as the drag force. Air that travels along the surface of the ball that is spinning in the opposite direction of the drag force travels slower.

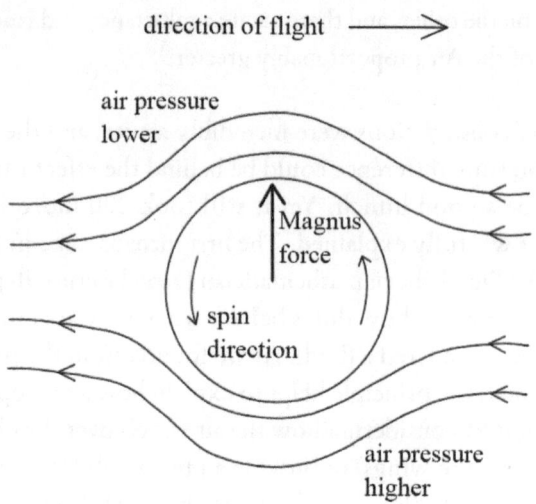

When a ball is spinning in an airflow, a Magnus force is generated.

The difference in speed on both sides creates a difference in pressure, which then results in a lifting force that is perpendicular to the drag.

When it comes to unstoppable shots in football, players often use aerodynamics – including the Magnus force – to trick opponents and score.* A hard straight kick will generate a lot of speed, which reduces the time the goalkeeper has to react, while a curving ball with a lot of side rotation (and speed) can take the ball around a four-person wall during a free-kick and into the corner of the goal to leave the goalkeeper bamboozled. The path of this trajectory is usually a circle, but in some cases it can be something much more extreme. A prime example is a goal by the Brazilian full-back Roberto Carlos against France in 1997. Perhaps one of the best goals ever scored in a competitive match, and certainly one of the most memorable shots in football history, the São Paulo-born defender was taking a free kick 35m away from the French goal. He carefully placed the ball down and then began to walk backwards, almost going to the halfway line. Once his rather long run-up was complete, he hit the ball hard and it was initially heading so far wide of the goal that a ball boy, who was standing a few metres to the right of the goal, ducked to get out of the way. But just as it seemed like the shot would be going wide, the ball began to curve strongly to the left and just snuck inside the goalpost. All the French goalkeeper, Fabian Bartez, could do was stand and watch, perplexed at what had just happened.†

---

* For different sports, the Magnus effect is greatest for a table tennis ball.

† For a video of the free kick, see youtu.be/crKwlbwvr88.

The jaw-dropping goal also sent physicists in a spin, and theories for what caused this huge swerve ranged from the material of the ball, the unusually dry conditions on the night to even a strong gust of wind. Yet Carlos knew what he was doing. He apparently spent hours practising the kick on the training ground and realised it was possible to produce such an extreme curvature. But doing so in a live match in front of tens of thousands of people was an altogether different matter. Some fourteen years later, in 2010, Quéré and colleagues investigated what could be going on with Carlos's shot. They experimented not with footballs, but instead with tiny polymer spheres. They fired such objects with a slingshot through water at different speeds and with varying spin rates. They then tracked the spheres' progress with a high-speed camera. The researchers witnessed that for high spins and speeds, a sphere undergoes a 'spinning-ball spiral' effect, where the friction exerted on the ball by its surroundings slows it down enough for the spin to take over in directing its trajectory.[6] In other words, at a certain point in its flight as it slows down, the influence from the rotation (spin) of the ball is much larger than the influence from the inertia of the ball. The result is that it stops following a circle and instead follows a spiral. 'The spiral is a fabulous trajectory,' Quéré told me. 'For a goalkeeper who is used to circles, a spiral would be much more vicious because it bends more and more.'

The researchers think that Carlos's shot was a perfect example of a 'spinning-ball spiral' that could only have happened under certain conditions. First, Carlos was able to hit the ball with such force to give it a starting velocity of about 30m/s (67mph) and with a lot of spin, which resulted in a strong Magnus force. But the third, and possibly most important, is that the ball was hit so far out from the goal that the ball had the chance to deaccelerate while still spinning quickly. This means that towards the end of the ball's flight, the spin took over and it started to follow a spinning-ball spiral, bringing it 'back' towards the goal. The team calculated that if the ball that Carlos hit went straight, rather than started to curve, it would have missed the goal by a mind-boggling 4m. They then transformed the kick into the arc of a circle, the usual trajectory of a free kick, finding it would miss the goal by 1m. 'This is what the goalkeeper likely anticipated,' adds Quéré. Indeed, that is certainly what the ball boy behind the goal expected. But by the ball's flight path transforming from a circle into a spiral at that last moment it gained one more metre and snuck inside the goalpost.* Remarkable. Quéré and colleagues calculate that the ideal spinning-ball spiral requires an initial velocity of 32m/s and imparting a spin on the ball of about 100 rotations per second, which not even Carlos could achieve. Yet no one has ever fully replicated that free kick in a match, probably because the chances of pulling it off are so low and the probability of getting it all wrong and kicking it into 'row Z' is very high.

That goal was an example of extreme spin, but there is another intriguing ball flight that can occur when a ball has very little or no spin. This is called a zigzagging knuckle-ball. The Portuguese footballer Cristiano Ronaldo has

---

* If gravity (and the goal) was absent to bring the ball down to Earth, the ball would have continued going around and around like a spiral, hence the name spinning-ball spiral.

made the knuckleball his go-to shot when taking a free kick, and he has scored rather a few of them during his career at Manchester United, Real Madrid, Juventus and Al-Nassr. In the case of the knuckleball, it is not so much a huge lateral movement of the ball that is surprising but the unexpected movement as it hovers in the air, zigzagging slightly from left to right as it travels. Indeed, the lateral deviation of a knuckleball may be on a par with the size of the ball itself – some 25 to 30cm – so one might think that it is easy to save. But many top goalkeepers have looked very silly in response to this kick, as they can only watch, feet planted to the ground, as the ball flies past them and into the goal.*

The first example of the impact of a knuckleball was not in football but in golf – a game that mixes long distances with precision, a tricky combination. In the sixteenth century, golf balls were made from hardwoods such as beech or boxroot and this caused people considerable anguish – even more so than in today's modern game. The balls were initially smooth, and players found that when they hit them far in the air, they wouldn't travel on straight paths but instead would oscillate, sometimes by as much as 10m from one side to the other. 'This meant golf at the time was unpredictable, more like a game than a sport,' adds Quéré. A further surprise came when the balls became old and scuffed: they actually performed better and didn't show this effect, travelling relatively straight. In the nineteenth century people

---

* During the 2010 FIFA World Cup in South Africa many goalkeepers complained that the official match ball – the Jabulani – was susceptible to moving in flight when kicked with little spin.

soon discovered that if they put some defects on to a new ball with a hammer, say, the balls would not only travel straighter but further. This was quite the finding given that one would expect a new ball to perform better than an old one, especially one that has been battered.

It turns out that the explanation for what was happening was related to a phenomenon in aerodynamics known as the 'drag crisis'. Usually, the drag force increases with the speed of the ball: the faster the ball moves, the higher the drag force to slow it down. But there is an interesting phenomenon that occurs at certain ball speeds:* the drag doesn't continue to rise but drops dramatically– a drag crisis. Some of the first clues about what was going on came in the late nineteenth century when in 1877 Lord Rayleigh (who examined the whistling kettle in Chapter 2) modelled the flow around a cylinder to describe the irregular flight of a rotating tennis ball. Rayleigh, who was also a keen tennis player, postulated that turbulent effects could be happening in the wake of a fast-moving ball.[7] The drag crisis itself, however, was first discovered in 1912 by the French civil engineer Gustave Eiffel, who built Paris's iconic tower. He found that when the speed of a sphere increased beyond a certain value, the magnitude of the drag reduced dramatically before levelling off as the speed increased further.

The drag crisis was explained a couple of years later by the German physicist Ludwig Prandtl, who is often referred to as the father of modern aerodynamics. And the

---

* I say speed here, but the proper term would be Reynolds number. Turbulence appears when this number, which is related to velocity and ball size, exceeds a certain value.

key behind it was what Rayleigh postulated: turbulence. When a ball is moving in the air, the air flows around it (as shown in the figure on page 146). This includes a layer of air at the surface of the ball called the 'boundary layer'. At low velocities the boundary layer is smooth, or laminar. Given that air cannot flow the whole way around the ball, the boundary layer separates at a certain point and for a smooth sphere it detaches relatively early from the surface. As it does so, eddies, or vortexes, form in the wake of the ball, but because it separates early the size of the wake behind the ball is large, resulting in a high drag. At higher velocities, however, the airflow at the surface of the sphere changes from being streamlined to turbulent and this turbulent air 'sticks' to the ball longer as it travels around and results in a later separation (see figure on page 157 showing how turbulent air can 'stick' to the ball). This results in a much smaller wake behind the ball compared to the case of laminar flow, and with it a reduced drag. This switch from laminar to turbulent flow is what is behind the drag crisis and what allows a ball to travel much further.

It turns out, however, that the drag crisis affects the motion of the ball in more than one way. At high velocities, the eddies in the wake are symmetric on both sides of the ball, meaning that it travels straight. Around the drag crisis, however, the eddies become asymmetric – different on one side of the ball than the other – and this creates a wake that 'tilts' either to the right or left in a random fashion. This then causes the path of the ball to wobble from side to side depending on the orientation of the wake. Coming back to golf, the smooth golf ball happened to be hit at a speed just where the drag crisis occurs, which resulted in it swaying from left to right in a random way, as was seen more than 100 years ago with wooden balls. Putting dimples on the surface of the ball, as modern balls have, results in adding turbulence to the boundary layer. This not only decreases the drag, allowing it to travel much further, but also puts it in a regime far away from the drag crisis so that it isn't affected by the wobble. In other words, the crisis now appears at much smaller velocities than the golf ball is usually travelling at, resulting in a straight trajectory. 'The dimples are an extraordinary thing that lets you play golf,' adds Quéré.

Quéré and colleagues wanted to work out, however, why a knuckleball can be seen in some sports, such as football, cricket, baseball and volleyball, but not for others, such as table tennis, squash or basketball. In 2016, they came up with an answer. The team used a 'kicking machine' to generate different ball speeds with very little spin. The machine included a motor connected to a 1.2m-long steel arm that had a flat 20 sq cm square plate that impacted the ball. They then tracked the motion of different balls using a high-speed camera.[8] For a relatively smooth football, the drag crisis occurs at a speed of about 25m/s, with the knuckleball flight happening at a shorter distance than the typical shooting range. This means it is possible to see the effect on the football pitch. Yet in other sports the opposite is the case and the distance for a ball to knuckle is much larger than a general shot distance, which is why it is never seen. Table tennis is one example where, despite the high speeds that a ping-pong ball can travel, its small size means you are too far below the regime where the drag crisis occurs. So while ping-pong balls are highly sensitive to air, at least they don't zigzag as they travel, which would make play very tricky.

Quéré and colleagues found that the knuckleball distance happens to be the shortest in baseball. This explains why the effect has been used since the early twentieth century when the Boston Red Sox pitcher Eddie Cicotte became such a master of the knuckleball pitch that he earned the nickname 'knuckles'. This type of pitch involves holding the ball between your thumb and knuckles (hence why it is called a knuckleball) so that when thrown it rotates slowly as it moves towards the batter. The knuckleball also must be launched at about 30m/s, to invoke the drag crisis, which is much slower compared to the usual baseball pitch of about 47m/s.

The pattern of stitches on a baseball also plays a role in affecting the flow of turbulent air around the ball, which explains why baseball is perhaps the ultimate knuckleball sport. Indeed, in 2021 researchers in the US found yet more surprising effects from a baseball knuckleball. It is not only difficult to throw a ball with little spin, but tiny changes in throwing technique – different release time or speed – can have huge effects on the outcome, making it unpredictable not only for the batter but the pitcher, too. This reliance on 'initial conditions' happens to be a key feature of chaos, which emerged in the early twentieth century. At that time, many scientists thought that the world was either deterministic or random – deterministic in the sense that it is regular and predictable while random events show no obvious patterns and are unpredictable. Yet science was thrown on its head when a third possibility emerged, chaos, in which apparently random behaviour in systems are governed by deterministic laws. Chaos was first discovered in weather forecasting and was later found in many other areas, such as population dynamics, simple pendulums and even in infant cries.*

Back to baseball, and one of the main features of a chaotic system is that any change to the initial conditions can have a drastic effect on the final outcome – commonly referred to as the butterfly effect. The researchers developed a simple model of the knuckleball and found that the motion of the ball is also chaotic: tiny changes in the initial orientation of the ball and its spin can result in huge difference in its final position by the time it reaches the batter. They modelled several pitches that each began with near-identical initial positions and velocities – the initial orientations and angular velocities varied by less than 0.01 per cent. Yet even with this tiny initial difference, it led to huge final outcomes for where the ball ended up, in some cases by as much as 1.2m, putting it even well outside the strike zone.[9]

So, while it is difficult to produce a knuckleball in baseball, predicting its trajectory is pretty much impossible and that is why it is so successful. For that you can thank chaos – and a century worth of progress in fluid dynamics.

As well as the Magnus force in a spinning ball and the knuckleball with no spin, cricket has developed into a sport that uses aerodynamics in perhaps more subtle ways. What sets a cricket ball apart from other balls is that it has about sixty to

---

* See Banks, M., *The Secret Science of Baby: The Surprising Physics of Creating a Human, from Conception to Birth – and Beyond* (Dallas, TX: BenBella Books, 2022).

eighty raised stitches arranged along six rows. The stitches lie along the equator and hold the two leather hemispheres together. However, given that the stitching on a cricket ball is more prominent than on a baseball, it can have a drastic effect on the dynamics of the ball. A new cricket ball is nice and shiny, and it is exactly this property that fast bowlers, and in particular swing bowlers, use to generate some interesting dynamics. The trick used by swing bowlers is to keep one side of the ball as smooth as possible while letting the other side get battered and rough. This is why you will see bowlers polishing one hemisphere with their white clothes (leaving a distinctive red stain, if they are playing with a red ball) or using spit or sweat, or a mixture of all three.

Bowlers use the seam and the surface smoothness to alter the behaviour of the flow of air around the ball to produce swing – a slight movement in the air from left to right or right to left. The slight deviation in the cricket ball's path can be enough to bamboozle batters so that they throw their bat towards the ball thinking they will make full contact, only for it to nick the edge and head straight to second slip to be caught. Conventional swing bowling involves pointing the seam towards the direction of the swing you want to generate with the rough side on that side (see figure opposite). Wind tunnel tests have shown that the maximum side force for this method of bowling is obtained when the ball is bowled at about 31m/s (so not at the top speed of a modern professional bowler) and with the seam angled at about 20° (zero being straight at the wicket).* At such speeds the flow

---

* See www.espncricinfo.com/story/the-science-of-swing-bowling-258645.

of air is laminar and on the smooth side of the ball it detaches before encountering the other seam, while on the other side it hits the seam turning turbulent and separates later. This creates a pressure difference between the two hemispheres so that the ball moves, or 'swings', in the direction of the seam.[*]

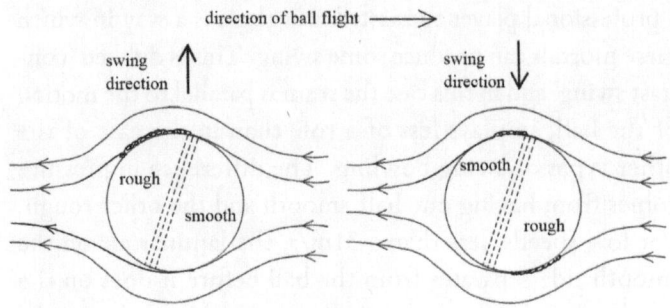

direction of ball flight ⟶

swing direction ↑

swing direction ↓

rough

smooth

smooth

rough

Swing bowling in cricket is about manipulating turbulent air via the seams and the roughness or smoothness of the ball's surface to produce swing (left) or reverse swing (right).

Bowling any quicker, as we found out earlier, can turn the boundary layer from being smooth to turbulent, so in this case both sides of the ball have turbulent flow at the boundary layer. But this can be useful, and top bowlers can use it to produce reverse swing, in which the swing direction is opposed to that of the seam direction. This happens when the speed of the ball is over 39m/s. In this case, on the smooth side, the turbulent air hits the seam but this time it is weakened and disrupted by the raised seam and so

* For a review, see Mehta, R.D., 'Aerodynamics of Sports Balls', *Annual Review of Fluid Mechanics*, vol. 17 (1985): 151–189.

detaches early. On the other rough side, the turbulent air separates later (and before it encounters the seam). This is enough to result in a lower pressure on the rough side and the ball swings away from the seam direction.

Keeping the seam at a certain angle and releasing the ball at certain speeds is no tall order and it takes years even for a professional player to master. But there is a way in which mere mortals can produce some swing. This is dubbed 'contrast swing' and in this case the seam is parallel to the motion of the ball, so plays less of a role than in the case of the other types of swing bowling. The difference in pressure comes from having one half smooth and the other rough. For low speeds, less than a 31m/s, the laminar air on the smooth side separates from the ball before it does on the rough side, where turbulent air is generated from the roughness of the ball and results in it sticking to the ball for longer. This produces swing towards the rough side. For quicker speeds, turbulence occurs on both sides, but the turbulent air is weakened on the rough side (in a similar way to how the seam in reverse swing weakens the turbulence), meaning that it detaches earlier from the ball than on the smooth side, resulting in a swing direction towards the smooth side – the opposite of what happened in the slower delivery. Contrast swing at slower speeds is easier for amateurs to bowl because it involves delivering a ball seam straight up and doesn't require great speed. Professionals, on the other hand, can use it to get the ball to swing in either direction just by bowling at different speeds.

Try that next time you are at the park.

8

UNDER THE WEATHER

NOBODY ENJOYS BEING ILL. That sense of impending doom when you detect a tickle on the back of your throat, your muscles are beginning to ache, and your eyes are a little heavier than usual. All of which usually signals that a cold, or worse, is about to hit. Unfortunately, those caring for young children will know the feeling all too well. When my two sons were in pre-school, a scratching sensation on the back of the throat would soon make a reappearance once summer was over and winter was just around the corner. A week didn't go by in the colder months without an e-mail warning parents and carers that a new bug was doing the rounds (and a reminder to keep children at home for two days minimum following the last bout of sickness or diarrhoea). I also learned at that time about diseases I never knew anything about before, such as impetigo and hand, foot and mouth, not to mention trying to identify the latest pimple or blemish that appeared on their skin from one week to the next.

The winter months are so much worse for illnesses because everyone tends to be confined indoors and all that mixing combined with little ventilation provides a fertile ground for the transmission of bugs. Many diseases, such as common colds, flu, measles and tuberculosis, spread via the respiratory tract. The virus or bacteria particles are held in liquid droplets that are then expelled by coughing or sneezing, with an unsuspecting victim inhaling these particles and becoming infected. The transmission of pathogens can also happen through direct person-to-person contact, such as a handshake or by contacting a contaminated surface and then touching your face, eyes, nose or mouth. This is why at the start of the COVID-19 pandemic in early 2020 we were all told to wash our hands as frequently as possible to stop the SARS-CoV-2 virus, which is responsible for COVID-19, from spreading (anyone wipe down their shopping/letters during COVID-19?). Information campaigns informed people how to wash their hands as if they had never done it before. One such poster from the World Health Organization, simply entitled 'how to handwash',[1] described a seven-step process that took between forty to sixty seconds to complete to make sure that the hands were 'safe'.

As people had much more time on their, er, hands due to lockdown, the poster was soon subjected to meme treatment. One example featured the handwashing images set to Lady Macbeth's fractured soliloquy 'Out damned spot' from Shakespeare's famous play.* Lady Macbeth washes her hands for 'a quarter of an hour' to cleanse her of a

---

* See www.penguin.co.uk/articles/2020/03/the-history-behind-the-lady-macbeth-coronavirus-meme.

murderous guilty conscience after compelling Macbeth to kill King Duncan. During COVID-19, we were told to wash our hands for at least twenty seconds – the amount of time that health officials deemed necessary to make sure they were thoroughly clean. And if you didn't want to count to twenty you could always, as the UK government told us, sing 'happy birthday' twice. People soon got tired of warbling that tune, especially when it wasn't anyone's birthday, and so, again, with time to burn, people soon came up with sections of other songs that you could hum or sing along to.

As we discovered in Chapter 3, soap is a surfactant. And these clever molecules contain two sections: one that likes water (called the head) and the other that doesn't (called the tail). When we wash our hands with soap and water, the tails of the soap molecule are attracted to areas where there is no water. When it comes across a virus particle, the surfactant molecule surrounds it before wedging the tail into the virus' outer layer to get to the middle (where there is no water). In doing so, it pops the virus like a balloon pops with a pin. The soap penetrates in, breaking the virus open and releasing the contents into the soapy water and rendering the virus impotent. This double action of surrounding the virus and then splitting it apart makes soap an incredibly powerful agent against some pathogens such as SARS-CoV-2.*

While washing hands and surfaces is one way to combat the spread of nasty bugs, it isn't enough on its own. We all

---

* While sanitising gel containing ethanol is also effective at breaking the virus apart, the main difference with soap and water is that a sanitiser isn't good at taking the virus away from the skin and may even end up leaving active virus on the hands. Despite this, sanitisers are still effective at rendering a virus ineffective when soap and water is unavailable.

know that many viruses, including SARS-CoV-2, travel through the air, and sitting next to someone sneezing means you could catch what they have, regardless of how clean your hands are (and they don't even have to show any symptoms given that SARS-CoV-2 spreads asymptomatically). This airborne route of transmission is something we have known since the turn of the twentieth century; however, before then the 'miasma theory' of disease transferral prevailed. This stated that infectious diseases were caught by breathing in noxious vapours produced via decomposing bodies. The idea that pathogens could propagate through the air was disregarded at the time when it was discovered that certain ailments like cholera or puerperal fever, which is caused by uterine infection following childbirth, could be contracted through water or via the hands.[2]

That began to change, however, when the German bacteriologist Carl Flügge discovered in 1897 that pathogens were present in droplets that were expelled by the respiratory tract. We don't tend to breathe in a virus directly, but rather through virus-laden fluid droplets that form in the respiratory tract of someone infected and are then expelled via the mouth and nose during breathing, coughing or sneezing. Flügge's 'droplet transmission' theory began to slowly gain traction, but it was still thought that such droplets could only travel a short distance thanks to being pulled down by gravity, like a projectile from a catapult (or a fern, see Chapter 5). The airborne transmission of diseases – that pathogens could take to the air and travel much further, for many metres – was widely rejected.

In the 1930s, the US scientist William Wells took on Flügge's mantle. He was studying tuberculosis transmission and, like Flügge, found that coughing and sneezing created large droplets. Yet he also discovered that they equally created smaller droplets. The size of a virus might range between 0.02 and 0.5 micrometres, but a virus-containing droplet can cover four orders of magnitude, with the smallest droplet being about 0.1 micrometres and the biggest 1 millimetre – about the size of the full stop at the end of this sentence. The larger ones settle faster and near the infected person, but the smaller droplets could evaporate when coming up against the colder air outside the body. This, he thought, created an aerosol that could suspend the particles in air for much longer and over a larger distance. Indeed, larger droplets are difficult to inhale directly as they are only influenced weakly by airflow (especially that created by someone breathing in). Smaller droplets, on the other hand, are potentially more infectious because they are more affected by airflow and so can easily be inhaled.* Wells's idea, however, was ridiculed at the time by the scientific community, which did not believe that an invisible cloud of particles could freely float in the air and – more importantly – do so for long enough to be breathed in by an unsuspecting victim.

In the 1950s, Wells and colleagues set out to finally prove that pathogens can indeed be airborne. At a military veteran's hospital in Baltimore, they built a chamber that was located above a six-bed tuberculosis ward and connected via an air vent. The team put 150 guinea pigs into the 'penthouse'

---

* There is a big debate in the scientific community around what size constitutes a drop and what size is an aerosol. A dividing size of 100 micrometres has been proposed because above this size particles are unlikely to be inhaled directly.

chamber (it sounds pleasant, but it really wasn't) so that they were directly exposed to the air from the patients in the ward. After a couple of years, the team found that, on average, about three guinea pigs per month became infected with tuberculosis – a number that agreed with the rate of infection that Wells had predicted. It seemed like unequivocal proof for airborne transmission. Yet some in the medical community were unconvinced. They argued that the guinea pigs could have been infected through food or water. The team went back to the drawing board and built a second chamber that held another group of 150 guinea pigs. This time, however, the air was continuously disinfected through ultraviolet radiation. In this case, contamination from the ward via the airborne route should be much less or even zero, but if it was indeed through another route such as water or food then the guinea pigs should still get infected at a similar rate as before. After a couple of years, Wells and colleagues found zero guinea pigs became infected with tuberculosis, finally proving the naysayers wrong.

Both studies were largely a success thanks to the meticulous work of Wells's colleague Cretyl Mills, who documented everything in incredible detail. This was done so well that she even knew where in the exposure chamber every infected guinea pig was housed and, by examining the drug resistance of the infecting organism, could even identify the specific person who infected the guinea pig – a lesson in how rigorous record-keeping can lead to findings beyond the initial design of the experiment.* The team

---

* Cretyl Mills was eventually infected with tuberculosis during the study. She died in 1990 aged 70.

published their results in 1962[3] but omitted Wells, who in the 1950s contracted a tumour in the spine and suffered significant mental health issues while in hospital.[4] He died a year after the study was published.

Despite this painstaking work, and the many careful studies that have since followed, there still exists a legacy of denial about the airborne nature of pathogens, which impacts our ability to tackle infectious diseases today.*

Two metres. It was a distance we all became accustomed to during the COVID-19 pandemic, whether it was how far to stand away from someone or the distance of those spots on the pavement outside supermarkets or in school playgrounds. This was the 'social distancing' gap to help reduce the transmission of the virus, and it originated from experiments and theory that considered the projectile motion of isolated drops moving through the air. Larger drops – those about 100 micrometres – that may contain more pathogens than smaller droplets, are indeed too heavy to stay airborne for more than a few seconds and so fall to the ground within about 1 or 2m of being expelled. Given this airborne nature, does 2m really help? I remember puzzling over the fact that people stood 2m apart behind each other in a queue, but the queues themselves were stationed about a metre next

---

to each other as if the virus could only move along a queue and not between.

It is thought that droplets up to 1 micrometre originate from the lower respiratory tract – the windpipe and the lungs. Larger droplets of about 300 micrometres in diameter[5] are produced in the upper respiratory tract, such as the mouth and nose.[6] A single sneeze can generate thousands, even tens of thousands, of droplets that have velocities up to 20m/s.[7] Coughing, on the other hand, produces ten to a hundred times fewer droplets than sneezing and with velocities of about 10m/s – but still Usain Bolt at top speed.* What gives droplets the ability to move long distances is being suspended by a cloud of moist air that when it meets the colder, drier air outside the body, rises. But this cloud seems to have its own dynamics, with the warm air not only helping to keep the cloud suspended but also stopping the droplets from evaporating. In 2016, researchers found that coughing and sneezing can produce turbulent gas clouds that travel up to 8m away from an infected person – much longer than a typical room in a house.[8] Not only that but the droplets inside can remain suspended in the air for several minutes, or even hours, depending on the temperature and humidity of the air.[9]

Most people are likely to recoil at someone sneezing or coughing in front of them and take some action, such as covering their mouth or nose, or turn their face away, but what about someone simply talking to you? Probably not in this case, unless you want to receive strange looks from

---

* For a close-up view of a sneeze filmed at 2,000 frames per second, see youtu.be/piCWFgwysu0.

whoever you are conversing with. Yet talking alone can expel droplets – regardless of what language is spoken[10] – with researchers finding that 'plosive', sounds such as 'p', can even lead to jet-like flows from the mouth. These can rapidly reach a metre in length and thirty seconds of conversation can stretch them to 2m.[11] A study in 2020 found that droplet clouds emitted during one minute of loud speech by an individual infected with the SARS-CoV-2 virus could contain more than 1,000 virus particles, with each second of speaking generating between twenty and ninety virus-containing droplets.[12] Talking for a long duration, say giving a presentation, can produce the same number of droplets as a single cough.* Researchers also found that even breathing can result in a fast-moving jet of carbon dioxide that extends over a metre. This cloud is also turbulent and so can transport droplets over a large distance – propelled across a small room in only a few seconds – and even reaching ceiling height.

So, what can help? Masks. Before the COVID-19 pandemic many of us would have never worn a mask (unless you work in healthcare or construction or have done certain types of home improvements, such as sanding floorboards). The first time I wore a mask was during a follow-up hospital appointment in early March 2020 after falling off my bike and breaking my collarbone a month earlier (not advised). Upon entering the hospital, I was handed a mask by the receptionist, but it was only afterwards that I realised I had been wearing it upside down the whole time. That

---

* For an online app that can calculate a safe time and occupancy limits for indoor spaces of various sizes, different age groups and whether people are wearing a mask or not, see indoor-covid-safety.herokuapp.com.

familiarity (and competence of mask-wearing) soon changed when we all became accustomed to donning a mask in an indoor setting. But their introduction wasn't immediate. In the UK masks were only made mandatory in shops in July 2020, some four months after the start of the pandemic. Like almost every aspect of the pandemic, masks became a very polarised issue, with some believing that they took away our liberties and that governments shouldn't mandate something we should be wearing. Others, meanwhile, claimed that wearing a mask wasn't a big deal, especially if it meant not only protecting yourself from getting COVID-19 but others, too. In some countries, such as Japan, for example, mask-wearing when ill (or otherwise) is commonplace all year round. Indeed, one of the difficulties with the COVID-19 pandemic – and why masks were such an effective measure – was the asymptomatic nature of the disease. Some people who had the virus did not display symptoms and so were unaware that they carried the virus and were potentially infecting others.

The simple reason why a well-fitted mask works is that it stops the airflow from the mouth and nose from escaping. In 2021, researchers in the US used an infrared imager to analyse how a mask stops the flow of air. They first measured the airflow of a person breathing in the absence of a mask, observing a fast-moving jet of carbon dioxide that extended over a metre. This jet was directed down towards the floor when the carbon dioxide was exhaled from the person's nose, as expected, and horizontally when it came from their mouth.[13] When the person wore a standard three-ply surgical mask, however, the team found that the airflow was directed upwards out of the top of the mask during gentle breathing, while only heavy breathing or shouting resulted in small jets of air penetrating the mask. Yet they only travelled around 10cm, still a vast improvement on the metre or more that happens when not wearing a mask.

The advantage of face masks is that they not only provide protection for the wearer by stopping virus-laden aerosolised particles that could otherwise be inhaled, but they also trap virus-laden droplets being expelled by an infected person. This 'two-way' effect gives mask-wearing some interesting mathematics. If you have a cloth mask that is, say, 50 per cent effective, it can still result in a 'total' effectiveness that is 75 per cent. This is because when two people wear a 50 per cent effective mask, it boosts the overall effectiveness. For example, if a contagious person is wearing a mask, then there is a 50 per cent chance of transmission, but if the other person is also wearing a mask, then this 50 per cent is reduced to 25 per cent for them to become ill, meaning a 75 per cent drop in transmission. Of course, not everyone will wear a mask, but even if only 50 per cent of the population wears a 50 per cent effective mask, it still results in a 44 per cent reduction in transmission.*

The 'R number' for the original variant of SARS-CoV-2 was estimated to be about 2.5. This is known as the basic reproduction number and is the average number of people one person with an infectious disease will likely infect in the future. An R number of 2.5, for example, would mean 100 infected people would likely go on to infect 250 others. During COVID-19, epidemiologists stated that it was necessary to get the R number below 1 for the pandemic to start to decline, or a reduction of about 60 per cent from 2.5. The

---

* For a great video explaining this, see youtu.be/Y47t9qLc9I4.

maths shows in this case you would only need 60 per cent of people wearing a mask that is about 60 per cent effective to make this happen.* That's the theory but what about the experimental evidence? A study in 2024 showed that wearing a face covering of some sort over not wearing a face mask reduced the chance of infection of the original 'delta' strain of SARS-CoV-2 by about 30 per cent in adults, although this number dropped when more infectious strains such as omicron came along.[14]

Once omicron emerged, we were all told that cloth masks were inadequate and only FFP2 (or N95 in the US) masks would work. Indeed, a cloth mask made of a woven material (which I admit I often used) is a terrible filter material, given the gaps between the cotton threads can be as large as 0.1mm, allowing droplets to go straight through.[15] So, what makes FFP2 so effective? You might think that a mask works like a strainer – the smaller the holes then the smaller the particles it lets through. But this is not the case, the gaps between the mesh fibres in the mask are much larger than the particles themselves. Masks work instead by getting the particles to stick to the fibres, a bit like a fly sticking to a particular strand on a spider's web, as we looked at in Chapter 5. At the micrometre scale, things are 'sticky' because of a weakly attractive electric force called the van der Waals force. Particles that are larger than 1 micrometre generally travel in straight lines and as the fibres in the mask make a mesh-like structure, the chances are that it will hit a fibre at some point. At the other end of the scale, small particles that are about 0.1 micrometre tend not to travel in straight

---

* To play around with the numbers, see aatishb.com/maskmath.

lines but move in random, zig-zag patterns, according to Brownian motion.* In a similar manner to large particles, this means they are also likely to eventually hit a fibre.

It turns out, however, that medium-sized particles of around a micrometre are the trickiest to catch. They are not affected so much by Brownian motion and don't travel in straight lines like larger particles; instead, they are carried on air currents that flow around the fibres. Surgical masks and cloth masks struggle to halt the transmission of this size of particle, but this is where masks like FFP2 excel.† There is still a lot that is unknown about how these types of masks work, but it is thought that during the manufacturing process the surface of the fibres obtain a distribution of negative and positive charges that generate an electric field. When the particles interact with this field, they become polarised and attracted to the fibre, a bit like being pulled by a tractor beam. The addition of this electric field attracts about ten times more particles compared to uncharged fibres and FFP2 masks can block about 95 per cent of particles, hence the name, N95 (for the US standard). A statistical physics study in 2024 by physicist Richard Sear from the University of Surrey, which used data from the UK National Health Service's COVID-19 app, found that if FFP2/N95 masks were adopted across the entire UK population it would have lowered the R number for COVID-19 transmission by a factor of nine, so an R rate of 2.5 would be reduced to about 0.3.[16]

---

* Named after the Scottish botanist Robert Brown, who first described the effect in 1827 when studying the movement of pollen in water.

† In the early 1990s, the Taiwanese-US materials scientist Peter Tsai invented the non-woven material that is now used to create N95 masks.

Not everyone may have agreed that wearing masks was necessary during the COVID-19 pandemic, and I admit it was a pain when you forgot your mask when out shopping, but there are other ways to stop transmission. Some places, such as schools, are difficult to mandate mask use, which is totally understandable for younger children. But improving ventilation and filtration can also help (including simply opening windows, which might not be that desirable in winter). Sear's work found that doubling the turnover of air in a room either through ventilation or air filtration systems could reduce transmission by about 30 per cent. 'Over the last thirty years, great work by engineers and physicists have given us almost perfect filtration materials for viruses, which make FFP2 and N95 masks at least 90 per cent effective,' Sear told me. 'We have the knowledge and materials to be better prepared for the next pandemic, we just have to use them.'

One of the stranger aspects of the COVID-19 pandemic included how wastewater in sewage systems turned out to be an effective early-warning system, able to spot signs of an outbreak before routine testing did so.* Before the pandemic, analysing wastewater in sewage was a relatively small field, mostly used to test for drug use, but when COVID-19 hit, examining wastewater for the virus became incredibly successful and useful for public health officials.

---

* COVIDPoops19 is a global dashboard for covid wastewater tracking; see arcg.is/1aummW.

It also highlighted that SAR-CoV-2 can be expelled from places other than just the mouth and nose. And it is not just coronaviruses but many other pathogens that can lurk in or near the toilet. Parents will be familiar with noroviruses, or winter vomiting bugs, especially in the bug-breeding grounds of pre-school and school. Its main symptoms include stomach cramps and diarrhoea and/or vomiting. In the US there are about 23 million norovirus infections each year, resulting in 50,000 hospitalisations. *Clostridium difficile*, or *C. difficile*, is responsible for 500,000 infections and 29,000 deaths each year in the US alone.

Splashes from surfaces, sinks or even toilet flushes are often not thought about much when it comes to spreading nasty bugs, but they should be. While it is known that surfaces in bathrooms, and especially the toilet, can be a breeding ground for bugs, not much was known about the aerosols that could be created even in the simple process of flushing the toilet. Experiments carried out in the 1950s in which cisterns were seeded with microorganisms found evidence for the blighters in the air and on nearby surfaces after the toilet was flushed. And even after multiple flushes, microorganisms still lurked in the toilet bowl. In 2020, researchers at Southeast University in Nanjing, China, carried out computational models of a flushing toilet bowl in common siphon toilets and looked at what happens not only to the liquid in the bowl itself, but the air around it. They found that as water pours into the bowl during a flush, it creates turbulence in the liquid that results in mini vortices. This makes sense given that there is a lot of noise when flushing a toilet and all that energy must go somewhere. The simulations show that these vortices kick out particles with upward velocities of about 5m/s and cause some 50 per cent of particles to even rise above the toilet seat itself, reaching a height of about 1m from the ground.[17] Even one minute after flushing, the researchers found that the particles were still at that height or similar, suspended in air.

But what about actual experiments? In 2022, John Crimaldi and colleagues from the University of Colorado in Boulder, Colorado, used lasers to illuminate the air surrounding a toilet and imaged and filmed the plumes that were thrown up when flushed.* The advantage of lasers is that you can use them to measure certain aspects of the plume without disrupting it. Crimaldi's speciality is using lasers to measure how objects are transported through fluids – how sea urchin sperm moves through water, for example. His research caught the attention of fellow Colorado engineers Karl Linden and Mark Hernandez, who work in areas such as wastewater treatment and disinfection. When COVID-19 emerged, they teamed up to investigate whether pathogens in faecal matter could be tossed into the air via flushing and how the droplets move. 'I wasn't really interested at first, but they talked me into it,' Crimaldi told me. 'But I am glad they did, as it turned out to be a really interesting problem and a fun project.'

The researchers started by filling a toilet with water, which had the seat down and no lid. When they flushed, they found a strong chaotic jet was created in the air just above the liquid, which then travels out of the toilet bowl at a speed of about 2m/s. While they thought that this would result in aerosol particles rising slowly, instead they

---

* For a video of the flushing, see youtu.be/aDIIhzc-FWg.

saw that they shot out like a rocket, reaching a height of 1.5m – about face height for some people – within eight seconds (ugh). Large droplets, those about 2 micrometres in diameter, rose over a metre in the air but generally fell to the ground quicker. Yet smaller droplets – 1 micrometre in diameter or smaller – went everywhere and remained in the air for several minutes, like what happens during a sneeze or cough, with some of the particles even reaching the ceiling.[18] 'All of us were caught off guard by just how energetic the plume was and how quickly it spread,' says Crimaldi. 'We were certainly surprised and shocked.'

One would think there is a simple solution to this problem: close the lid before flushing, as we all try to do, honestly. But(t) think again. In 2024, Charles Gerba from the University of Arizona, who first got interested in the aerosols from toilets in 1973, lifted the lid on closing the lid by seeding toilets with the bacteria *E.coli* as well as with the bacteriophage MS2, which is a virus that infects *E.coli*. Gerba and colleagues tested two types: a lidless toilet located in an office building that had a U-shaped seat, as well as a standard toilet with a lid that you might find in a typical home. A minute after flushing the toilets, they took samples from the floor, walls, toilet seat and the rim of the toilet bowl. They found that regardless of whether the seat was down or up, the level of contamination in the toilet and around the room was similar.[19] For the home toilet, putting the lid down, however, resulted in *more* contamination on the floor and the walls compared with having the seat up. This is because closing the seat results in the jet plumes being sprayed towards the floor. After all, the lid is not airtight.

So, what can you do? Of course, even if some faecal matter is expelled from the toilet during each flush, it doesn't mean that you are going to get ill each time. Otherwise, we would never be well. Similarly, if someone is coughing on the bus near you it doesn't mean you need to book the next week off work. But it won't come as a surprise that rigorous cleaning can help to reduce the chance of faecal-oral transmission in the bathroom. Adding a disinfectant to the toilet bowl before flushing as well as using a detergent dispenser in the cistern reduces contamination everywhere, especially in the bowl water itself. Compared to just cleaning the bowl with a toilet brush, adding a disinfectant reduced contamination by 99.99 per cent. Gerba and colleagues suggest that when someone in the household is ill with norovirus then the toilet bowl, seat and lid should be disinfected regularly together with the floor. Gerba says in general it is best to clean the toilet bowl every three to four days so as not to allow a 'bacterial biofilm' in the toilet bowl to grow too much. 'I tend to flush and run based on our results,' Gerba told me. And another thing: you might want to keep your toothbrush as far away from the toilet as you can, otherwise you are effectively brushing your teeth with what has been in the toilet.

Gerba and colleagues are now planning to study the impact of different toilets around the world as well as different flushing mechanisms, which he says could make a 'big difference' in the release of aerosols. 'I always tell people my career has always been in the toilet,' adds Gerba, 'and I now have several research papers to prove it.'

FRIENDS AND FAMILY

HOW MANY FRIENDS DO you have? Ten? Twenty? Perhaps a couple of best friends and ten other people who you would call close? How would you deem someone close as opposed to just a friend or an acquaintance? As we go about our business each day we talk or communicate to many people – friends, colleagues, family, neighbours and those we don't know. Humans are social animals, after all, and throughout life we develop and maintain relationships with many different people. First with our parents and siblings, later with friends and then a partner, before maybe with children and grandchildren. These social relationships reflect a dependence on, and investment in, family and friends. According to the Oxford University anthropologist and evolutionary psychologist Robin Dunbar, we all build up a hierarchy of social relationships. In the 1990s he proposed a system that featured layers of friendship, or what is now known as Dunbar's number. He did so after discovering a correlation between the size of primates' brains and the size of their social group. When extrapolating the finding to humans, Dunbar postulated that the maximum number of people an adult human can maintain a social connection with is approximately 150, which includes both friends and family members.[1] Dunbar's theory suggests that maintaining relationships beyond this number becomes increasingly difficult due to our brain's limited ability to keep all these relations going – and even modern technology, such as apps, don't help much either.

Dunbar postulated that these 150 people are organised in a series of five circles that go from small, more intense relationships to larger, less-intensive associations (see figure opposite). Each consecutive circle has roughly three times the number of individuals as the circle inside it. The innermost layer is called the 'support clique', which consists of about three to five people who are usually immediate family members or very close friends. The next layer expands to the 'sympathy group', with up to ten additional people, including extended family and close friends. Moving further outward is the 'bonded group', which makes up some thirty-five individuals consisting of friends, colleagues and neighbours. The outermost circle is known as the 'tribe', consisting of 100 people, including acquaintances and casual friends. The size of the Dunbar circles is assumed to remain stable throughout life, even though the individuals in our lives may change. It is thought that the support clique of five members, which usually consist of parents, spouse and siblings, receives about 40 per cent of our 'social effort', with the ten additional members of the sympathy group getting about 20 per cent. The remaining 135 individuals share the left-over 40 per cent.[2]

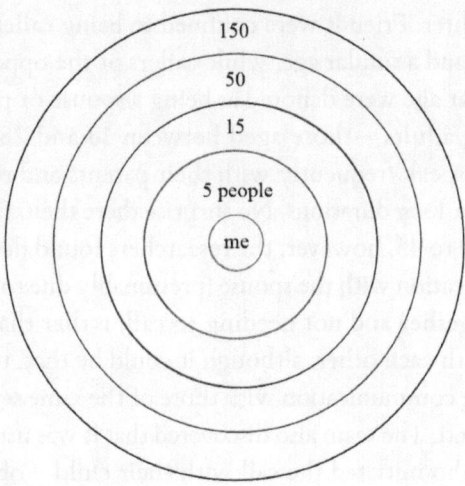

150

50

15

5 people

me

Dunbar's layer of friendships.

In 2016, Tamás Dávid-Barrett and colleagues from the University of Oxford wanted to know in what way relationships shift over an adult's life. To do so, they analysed mobile phone data, which included over 3 billion calls made in 2007 in a single European country. This was a year when people, and this might come to a shock for anyone under the age of 30, mostly communicated via phone. It was also a time when the proliferation of smartphones and apps like WhatsApp were just about to take off (although texting was still a big thing back then). The dataset included call frequency, who initiated the call, its duration, as well as metadata on the caller's age and gender. From this they identified 2.5 million men and 1.8 million women, and by looking at the age and gender they deduced certain relationships. An age gap of twenty-five years, for example, represented a family generation, such as between a mother and daughter. Friends were confined to being callers of the same sex and a similar age, while callers of the opposite sex and similar age were denoted as being a spouse or partner.

Young adults – those aged between 18 and 28 – were found to speak frequently with their parents and romantic partner for long durations. No surprise there then. From the ages of 29 to 45, however, the researchers found decreasing communication with the spouse (presumably due to the pair living together and not needing to call, rather than being fed up with each other, although it could be that, too), but increasing communication with those of the same sex, likely a best friend. The team also discovered that it was usually the parents who initiated the call with their child – obviously wanting to know what they were up to – but when a person reached their mid-thirties, they were more likely to initiate the phone call with their parents (and daughters were more likely to call their mother than their father). This could be a sign that as they became parents themselves, they were seeking more support from their parents, and who could blame them. Between 46 to 55 years old, people were balancing communication across three generations: their children, their friends and spouse, and their parents. Yet the study found that this 'cross-generational' communication was mostly done by women – a sign, perhaps, that women do more to hold together different generations of the family than men do.[3]

Admittedly, there was a fair amount of generalisation in the analysis, especially, for example, assuming that calls between sexes are friends rather than partners. So, in 2023 the researchers instead analysed almost 4 billion mobile phone calls made in 2015 in Chile. The key difference was that this data provided individual surnames.* Chileans usually inherit two surnames: the first is usually the family name of the father and the second is the first family name of the mother. In this case the researchers could clearly differentiate callers from being friends or family. They discovered that call frequency, length and share among all calls is higher for relatives than it is for friends.[4] As found in the previous study, when a person is in their early twenties, it is usually the mother who initiates the call (but the calls tend to be short). When in their late twenties, however, it is the child who calls their mother and the length of the call doubles, which the researchers put down to 'grandmaternal support'.† Both men and women are also more likely to conduct frequent and longer calls with their mother than their father. The study mainly found that kinship remains strong regardless of issues such as migration, which can reduce the physical contact between family members. Indeed, the relationship between family members and friends is different; after all, you can choose your friends (to some extent) but you are stuck with your family.

When it comes to friends, there is no greater turmoil than during the teenage years – changing school, interests, and suffering from raging hormones. In 2023, researchers from Spain and the UK examined how friendships changed among 221 Spanish teenagers during two consecutive academic years from 2020 to 2022.[5] They conducted surveys with the students every three to five months to gauge their

---

* The actual surnames were hashed to comply with data protection regulations.

† The mid-twenties being the mean age at which women have their first child in Chile.

friendship circles of five best friends and a further ten close friends. While it is thought that friendships during the teenage years can be subject to a fair amount of turmoil, perhaps changing by up to 30 per cent of friendships each year,[6] the study found that the number of friendships and best friendships stayed remarkably consistent across the two years. But that doesn't mean there weren't changes. Best friends are usually stable but if someone moves out of the best friend circle, they usually remain a close friend. For those in the close friend circle, however, this group is more dynamic, and people can completely disappear (not literally, but just from the person's list of friends) or in some cases can enter the best friend circle. This goes some way to show that the intensity of relationships in the first circle is stronger than the second one, given that it is very rare for a best friend to disappear completely.

The study also found that one of the key drivers of friendships and where the strongest friendships develop is by being in the same school class, with best friendships among students who had never shared a class being rare. When a student changes class this can also result in a loss of a sizable fraction of friendships, but it is not all doom and gloom, as new friendships can form in the new class. The researchers even suggest a nice physics-based metaphor of friendships being like a 'social atom'. The layers play the role of atomic orbitals, and individuals are electrons that sit in the orbitals. The inner-most orbitals (being the five best friends and ten friends) attract electrons more strongly than outer ones, so there is less turnover. Yet electrons may leave their orbitals for good, leaving a 'hole' that soon gets filled by another electron. This also shows how frequent interactions are important when maintaining close bonds among friends – more so than it is with family members. After all, you might only see your uncle once a year at the Christmas table, but after a while it will feel like you have hardly been separated. Just remember to keep laughing at the bad jokes.

That's all well and good for the raging hormonal period of the teenage years, but what about the turnover of friendships throughout life? A study of mobile phone calls made between 2007 and 2009 from a southern European country took the top five people in a person's call network – seeing these as likely to be in the top two Dunbar layers of social relationships – and how the call length and frequency changed over time. They split callers into three groups: young adults (aged between 17 and 21), adults (aged between 25 and 35) and middle-aged (aged between 45 and 55). For same-gender friendships, they found that both male and female adults and young adults made new friends as quickly as they lost friends, with young adults having a higher turnover of friends over adults and middle-aged adults.[*] But things changed when it came to opposite-gender friendships. In young adults, they found that women tended to lose opposite-gender relationships over time without replacing them, while men seemed to make opposite-gender friends as easily as they lost them.[7] This could be a sign that women were choosier when it came to male friends compared to men with female friends. By middle age, however, these friendships had settled down as

---

[*] Both female and male adults lose opposite-gender friendships quicker than they form new ones (although the percentage here is small, representing under 5 per cent of friends). This could be due to significant life events, such as getting married or having children, which can cause a significant upheaval in friendship circles.

people started a family and had a stable career (well, hopefully). In general, the study found that women tended to have fewer but closer relationships than men, which included two very close people in their lives – a partner or spouse and a best friend. On the other hand, men tended to prioritise the women in their lives, such as spouse or partner and mother, over their male friends. And men usually had one 'special' friend, either a romantic partner or a male friend, but not at the same time. All of which goes to show that one of the biggest upheavals of the social hierarchy in adulthood is the dating game, so let's have a look at what statistical physics can reveal when it comes to courtship.

Meeting people before the 2000s would have been the preserve of going out (shock!) to clubs or bars, or perhaps via hobby groups and other social activities. Today, however, it can be done sat on the sofa in your pyjamas with a flick of the finger using a dating app on a mobile phone.[*] According to the US-based thinktank Pew Research Center, 30 per cent of US adults in 2023 said that they have used a dating site or app in the past – a percentage that rises to 50 per cent for those aged up to 30 (but just 20 per cent for those aged between 50 and 64).[8] Some 42 per cent of adults in the Pew survey said dating apps made it easier to find a partner, with 32 per cent saying it made no difference and 22 per cent claiming it was harder.

---

[*] There could be as many as 1,500 dating apps; see financesonline.com/online-dating-statistics.

In general, couples tend to resemble each other in terms of age, interests, education and physical attractiveness. An explanation for this is the so-called 'matching hypothesis', in which men and women are both attracted to those who look and do the same things as them. Another school of thought is the 'competition hypothesis', which presumes that there is a consensus about what makes a desirable partner; therefore, people pursue those who are deemed most desirable, regardless of their own desirability. This can, however, also produce the same effect as the 'matching hypothesis' in that couples who are the most desirable tend to pair together, followed by the next desirable, and so on. While these two hypotheses can result in the same outcomes, it is difficult to know whether there is a consensus around desirability and what traits people tend to pursue.

Well that was the case before all these dating apps came along, which collected vast amounts of data that scientists could mine and analyse via statistical techniques, just as with the case of mobile phone usage. In 2018, researchers in the US examined messages that were sent between adult heterosexuals on a US dating app over a one-month period in 2014 in four cities: New York, Boston, Chicago and Seattle.[9] Perhaps unsurprisingly, women received more messages than men, but a small fraction of the population – both men and women – received a lot more messages than the average. This was remarkably consistent across all four cities. The single demographic that received the most messages – at a rate of one every thirty minutes – was a 30-year-old woman in New York. This message onslaught for women tallies with the Pew data, in which 54 per cent of women said they were overwhelmed by the number of messages they received, while only a quarter of men had such a complaint.

The researchers employed an algorithm akin to 'PageRank', which Google uses to calculate the rank of web pages in their search-engine results, to rank the population in each city. It used messaging data (not only the number of messages but the type of person who contacted you) to position men and women from least to most desirable – a score of one being most desirable and zero being the least. After calculating the desirability score, the algorithm looked at various attributes such as age and education to pick out any trends. For women, it found that desirability slowly falls with age, being highest at around 25 years old and lowest at 60. The reverse, however, happens for men. Desirability is lowest for those in their twenties, peaking for men in their fifties before then declining. Education was another difference among genders. For women, the more educated a man, the more desirable, but for men, the sweet spot was a woman with a college education. Desirability was lower for both a school education and a postgraduate degree.

The researchers then scrutinised communication between individuals based on their desirability. If the least desirable man sent a message to the most desirable woman it resulted in a desirability communication gap of +1. Likewise, if the most desirable man sent a message to the least desirable woman, then the gap would be -1. Both men and women commonly contacted those who had a similar ranking to themselves, despite people not having access to their own desirability score. This indicates that most people have some sense of where they sit in the 'desirability spectra' of a population.* For both sexes, however, there was a significant tendency to contact partners who were deemed by the algorithm to be more desirable than themselves – contacting those around 25 per cent, on average, further up the rankings. For men, it is the norm to send messages to women who are more desirable than themselves, whereas very few men contacted women who were deemed by the algorithm to be significantly less desirable. Yet there is a significant difference between genders in how those messages are composed. Both men and women wrote longer messages – sometimes twice as long – to people considered more desirable than themselves.† Women also used a greater percentage of 'positive' words in a message when communicating with those deemed more desirable. Men, for some reason, do the opposite and use fewer optimistic words.

The question most people might be wondering is what this data can tell us about whether punching above one's weight – to coin a term from boxing, perhaps not the most apt – works? When looking at the desirability gap in terms of receiving replies, the probability of receiving a reply declines the larger the desirability gap, as one might expect.

---

* A reanalysis of a landmark 'meta-analysis' carried out in 1988, which gathered data on twenty-seven different studies of attractiveness and couples, found that we tend to date and marry people who are similar when it comes to physical attractiveness. See Webster, G.D., Li., Z., Park, S.Y., et al., 'Dyadic Secondary Meta-analysis Attractiveness in Mixed-Sex Couples', *Personality and Individual Differences*, vol. 228 (2024): 112730.

† These trends were similar across all four cities, except in one aspect: message length. For some reason, men in Seattle wrote almost double that of men in the three other cities – perhaps they are more literary?

Women are more likely than men to receive a reply from someone deemed more desirable – the reply rate for men sending messages to more desirable women is about 20 per cent. Men are more than twice as likely to receive a reply from a woman less desirable than themselves than from someone more desirable. The bottom line for men is that if you want a reply then go for someone who might have a similar desirability level. And if you must try to attract the attention of someone who might be considered way 'out of your league', don't expect a reply.

One app that works a bit differently than just getting in touch with loads of people and seeing who might reply is Tinder. Launched in 2012, it is one of the most popular dating apps, especially among those under 30 years of age. According to Pew Research Center, of those that use online apps, almost half use Tinder, rising to 79 per cent of US dating users aged between 19 and 29. In contrast, users aged 50 and older are about five times more likely to use Match (which is the second most popular online dating site in the US) than Tinder (only 11 per cent of those aged 50 and older used Tinder). Tinder works – I have been told – by showing users pictures of people who live nearby, whom they can either 'like' (by swiping right on the app) or 'dislike' (by swiping left) based on first impressions. If two users 'like' each other this is considered a match and they are allowed to communicate on the app.

To investigate how men and women use the app, in 2016 researchers created (for research purposes, obviously) fourteen fake profiles of what might be deemed 'average' people who live in London and New York. The profiles only used facial shots to prevent users from deducing information about income or even education. Tapping into data on half a million users, they found that male profiles tend to slowly build up matches over time, while women gain matches rapidly – more than 200 in the first hour.[10] Yet women tend to be highly selective in who they like, while men aren't so picky, swiping right far more often than women do. This results in a feedback loop, where men are less selective in the hope of getting a match, while women know that if they like someone there is a decent chance it will likely result in a match. Indeed, the study found that men had a matching rate of 0.6 per cent compared to 10.5 per cent for women.

Once matched, some 20 per cent of women send a message to the other person, but, surprisingly, only 7 per cent of men do. This might suggest that women are more engaged than men about potential matches, perhaps because they have been more careful in who they selected in the first place rather than the more scattergun approach by men. When men do message, they do so quickly. Some 60 per cent of men reply within five minutes of a match notification, whereas women tend to wait (only 18 per cent sent a message within five minutes). The reason for this rapid communication could be because men write short messages – about twelve characters long on average – so it doesn't take them long to reply. A quarter of messages from men were even fewer than six characters long, presumably 'Hello' or even 'Hi'. Women, on the other hand, pen longer first-contact messages, some well over 100 characters.

That's all well and good, but how can all this statistical analysis improve your chances in the dating game? The answer could lie in pictures. Women who increased the number of profile images from one to three resulted in a

37 per cent increase in matches. This jump was even bigger for men, where a profile with a single picture had only forty-four matches after four hours, but the number of matches increased to 238 with three images. And it is not just images that can help. While bios are short on Tinder, providing little or no information, men without any biographical information received an average of sixteen matches from women, but this increased to sixty-nine with a bio. And it doesn't even have to be long – something short such as 'Hello, I'm from London' is good enough.

See how easy people have it these days.

Slightly changing tack now, but once that match has happened and you decide to meet up, perhaps going on a few dates, things start to move quickly and perhaps, just perhaps, love is in the air. But what happens when you start to develop feelings for someone? In 2022, in one of the most left-field physics papers seen for a while, Dmitry Solnyshkov and Guillaume Malpuech from the Pascal Institute at the University of Clermont, France, postulated that courtship and love may be the result of a certain 'transition' in the brain. The idea to study the physics of love apparently came to Solnyshkov during a dream. 'I woke up and realised that this idea should be true,' he told me. 'I was surprised and excited when I had it.' The problem faced by Solnyshkov and Malpuech was how to study this in practice, given it is not possible to analyse human feelings directly. Instead, the authors did something that you don't often see in physics publications: they examined works of fiction for descriptions of love.

The duo chose three works: William Shakespeare's *Romeo and Juliet*, Honoré de Balzac's *The Lily of the Valley* and Jack London's *Martin Eden*. For each book, they took consecutive phrases that describes to the reader how a character feels about another. In *Romeo and Juliet*, for example, phrases include: 'What lady's that, which doth enrich the hand of yonder knight?'; 'O, she doth teach the torches to burn bright!'; and 'she hangs upon the cheek of night Like a rich jewel in an Ethiop's ear, beauty too rich for use'. Or from *Martin Eden*: 'He did not know how she was dressed, except that the dress was as wonderful as she'; 'He likened her to a pale gold flower upon a slender stem'; 'No, she was a spirit, a divinity, a goddess'; as well as 'such sublimated beauty was not of the earth'.

You don't see those kinds of sentences on Tinder.

Solnyshkov and Malpuech then asked people to score these phrases on a scale of zero to ten for intensity of feelings: zero signalling no feelings and ten representing full-on obsession. When they examined how such 'feelings' developed over time, it had a striking similarity with the theory of 'phase transitions'. Bear with me on this one. A phase transition in everyday life is what happens when water turns into ice. As you reduce the temperature of the liquid water, some parts of it – such as the surface or interior – start to freeze but the whole volume of water is not yet frozen. Then, suddenly, the whole volume freezes – what in statistical physics is called a phases transition. A phase transition is often associated with ordering – disordered water molecules become ordered in ice, for example. An important point about phase transitions is the order parameter. This is a measure of the degree of order across the 'boundary' of the phase transition. So, when a liquid is cooled to become a solid, at the freezing point the order parameter jumps from zero to a non-zero value before then saturating when the whole body transitions into a new state.*

I hope you are still reading. When Solnyshkov and Malpuech plotted the 'feelings for a character' as determined by the test readers against 'time', or when the romantic passages appeared in the narrative of the book, they found that there is indeed a quick growth of feeling, before it reaches saturation, just like an order parameter in a phase transition. And there is more. Some think that the brain operates near the 'critical point' of a phase transition, and by that I don't mean between a solid and a liquid where the brain just turns to yuck. But this criticality, it is suggested, allows the brain to achieve efficiency, ensuring the right balance between excitation and inhibition. In other words, a small change, known as a perturbation in physics parlance, can result in a large change overall. Levels of hormones in the brain, such as dopamine and serotonin,† would be responsible for moving away from this tipping point, with the brain switching from its normal operation to a supercritical regime because of the increase in excitation (more dopamine), for example. The intensity of feelings, therefore, could be related to the order parameter: initially a person may have little or no feelings for who they are dating, but then during courtship those

---

* A discontinuous jump is known as a first-order phase transition, while a more gradual change from zero is known as a second-order phase transition.

† Dopamine results in feelings of pleasure and motivation, while serotonin acts to control mood.

feelings begin to emerge, somewhat like an order parameter emerging to signal order from disorder.[11]

Solnyshkov and Malpuech postulate that if the brain does cross into supercriticality then it could be what makes a person 'blind' to the defects of their loved one as well as deaf to the warnings of others. The infinitely fast growth of feelings predicted by the order parameter theory, for example, could be responsible for the saying 'love at first sight'. But they also warn that this crossing could be random, happening at any moment. 'We see new faces all the time, but when the person becomes "ready" and crosses the transition point, the new face seen at this particular moment becomes the object of spontaneous symmetry breaking, which means that this effect is essentially random,' the researchers write. In other words, someone could fall for anyone at that particular moment. The authors say that perhaps Juliet seemed exceptional to Romeo, for example, not because she was actually exceptional, but because he just happened to have crossed the transition point inside himself when he saw her.

Talk about putting a dampener on one of the greatest love stories ever written.

The researchers also went about testing progressive love rather than instantaneous love. They again turned to fiction to do so, this time examining Charlotte Brontë's 1847 novel *Jane Eyre*. They did so not only because there is a delay between the initial meeting between Jane Eyre and Edward Rochester (referred to as Mr Rochester) and the several weeks in the timeline of the book before they fall in love, but also because it is possible to date the paragraphs accurately in the narrative. For example, on the seventeenth day since Jane first met Mr Rochester she says: 'I am sure most people would have thought him an ugly man; yet there was so much unconscious pride in his port; so much ease in his demeanour.' But then on day twenty-five Jane declares: 'I felt at times as if he were my relation rather than my master. So happy, so gratified did I become with this new interest.' Then finally on day forty-five: 'I had not intended to love him; the reader knows I had wrought hard to extirpate from my soul the germs of love there detected; and now, at the first renewed view of him, they spontaneously arrived, green and strong! He made me love him without looking at me.'

By studying the strength of the language used over time, again they found it agreed with the theory of phase transitions and day twenty-eight was found to be the origin of the 'transition'. When love is born from a strong liking or friendship, it does not cause the same level of intensity of feelings at the transition than 'love at first sight'. Yet the authors say this is not necessarily a bad thing. 'People who do not fall in love at first sight and [instead] feel a smooth increase of their feelings ... should not worry too much about it,' the authors write. 'While being difficult to distinguish from friendship, their relationship is actually more secure precisely because of this.'

Solnyshkov even notes that the theory could be used to predict the stability of a marriage. 'If one analyses a private diary and assigns a mark to each note describing an episode of falling in love, it might really have some predictive power for the stability of a future marriage,' says Solnyshkov. 'If the theory is true, of course'.

Many marriages, sadly, end by people going their separate ways, with the overall divorce rate in the UK being about 32 per cent.[12] Admittedly, this chapter has been an emotional roller coaster, starting off with calling friends on the phone, then marriage and now separation. While divorce is a complicated and stressful situation, regardless of the circumstances, it is even more so when children are involved. Many custody arrangements instruct each parent to have the kids every other weekend, but what happens when divorced parents have children with another, or more, ex-partner? The problem is trying to get all the siblings together on the same weekend when a person may also be in a relationship with someone who has kids with someone else. Is it still possible to find a custody arrangement where parents see all their children together every other weekend?

In 2014, researchers in Chile attempted to provide an answer. They modelled all the possible permutations from a theoretical perspective to see if it was possible to have all the kids every other weekend. Using graph theory, they denoted men and women as nodes with those who had children together represented by a connection between the two nodes. On these links, they then put an arrow. If it pointed towards a node, it meant that they had the children that weekend, while an arrow pointing away meant the children were with their ex-partner. They discovered that attempting to find a solution to satisfy everyone had an analogy in condensed-matter physics. In this case it was akin to finding the lowest energy state of a 'spin-glass' system – a diluted magnetic material, in which the 'spins' of the magnetic atoms are interacting randomly. In such a model it can be difficult to find an equilibrium state where all the spins are 'happy' in their configuration. And in a similar manner, the researchers concluded that such custody arrangements where everyone was happy – i.e., they either had all the kids on a weekend or they had none – was not possible.

That's the theory, but what about the reality? The team next modelled the mathematical rules on to real-life data. They ran a simulation with 10,000 'people' and assigned 16 per cent of women as having no kids and 20 per cent of both men and women as having kids with more than one partner. They assumed that 90 per cent of people were in a current relationship and that there was an 85 per cent chance that they had a child with that partner. When 20 per cent of the population have children with more than one partner then the theory found it could be possible to find a happy solution for all. Things become trickier when the population with children from more than one partner nears 30 per cent (which is not uncommon in some communities). But, even then, it is not all doom and gloom. The researchers discovered that it is possible to have an arrangement in which one of the parents can see all of their children every other weekend.[13] This so-called 'convenient' state can be created where all individuals have their respective children with them every other weekend, but some couples may not have all their children together. In truth, one might think this is the more realistic solution; after all, who has that number of beds to accommodate potentially so many children anyway.

Of course, life is more complicated than just thinking about nodes on a graph, with issues such as individuals having to work certain weekends or that everyone might not be willing to co-operate and communicate in an open manner. Perhaps sitting down with a diary, rather than a graph–theory analysis, and coming up with something reasonable may be the best bet.

10

THRILLS AND SPILLS

HEARING THE CHARACTERISTIC 'POP' of a champagne bottle as it is uncorked is one of life's great delights. A bottle of sparkling wine is a symbol of pure decadence, with champagne and celebration going together like bacon and eggs, Batman and Robin, or salt and pepper. There has only been one time in my life (so far, at least) where I have indulged in an expensive bottle of champagne. It was the COVID-19 lockdown in early 2021 and it was my wife's birthday (a special number, I won't say which, otherwise I might need to re-read that physics of divorce study more carefully). As it wasn't possible to book a fancy restaurant or a retreat to some far-flung location, or go anywhere for that matter, celebrations were confined to the house. I lined up a few activities and splashed out on a 2010 bottle of Dom Pérignon for, er, both of us to enjoy in the evening when the kids had gone to bed. As I opened the champagne, I was slightly nervous, not sure what to expect. After all, I would usually be uncorking a £15 bottle, or maybe £20–£30 if we really treated ourselves.

Would something costing ten times more be ten times better? The answer: yes. One of the main differences was the explosion of tiny bubbles as the champagne hit the tongue. It was an incredible sensation – a bite of carbonation that delivered copious amounts of texture and flavour as the bubbles popped in the mouth. If I had the money, I wouldn't hesitate to splurge on another bottle, for the right occasion, of course.

Making champagne is an art form that has been refined for centuries. The old myth is that the Benedictine monk Dom Pierre Pérignon discovered the method in 1697, but the production of sparkling wine was first documented in 1662 in England – some thirty years before Pérignon's feat. Yet it was only in the nineteenth century that production began in earnest in the Champagne region of France. Champagne is made by fermenting grapes in open vats before adding sugar and yeast to begin a second stage of fermentation, which takes place once the wine has been transferred to bottles that are then sealed. As the microorganisms consume glucose and fructose in the grape juice, they convert the sugar molecules into ethanol and carbon dioxide, or CO_2, which dissolves in the wine. The resulting champagne is a mixture of water and ethanol that is supersaturated* with dissolved carbon dioxide. This process of supersaturation was first described by the nineteenth-century English chemist William Henry in what is now known as Henry's law. It states that the concentration of dissolved gas is proportional to the pressure in the 'headspace' above the liquid. So, the more pressure between the wine and the cork the more carbon dioxide is dissolved.

---

* Supersaturated is when a solution contains more 'solute' than it can normally hold when in a stable state.

The pressure in the headspace in a chilled champagne bottle is about 5 bar – more than twice that found inside a typical car tyre – while the pressure in a bottle at room temperature is even more, about 7 bar (the warmer the gas, the quicker the molecules in it move and the greater the pressure on the container walls). This is why champagne not only comes in bottles with tough glass but also with a tight-fitting cork, which is compressed before it is inserted into the bottle to seal it.* In an unopened bottle, the dissolved carbon dioxide is in balance with the gas in the headspace, so there aren't that many bubbles in the wine itself. Uncorking the bottle, however, throws off this chemical equilibrium and to reach a new balance the dissolved carbon dioxide gushes out as bubbles in the liquid. For a standard 75cl bottle, the amount of carbon dioxide released is about six times the volume of the bottle itself.

Thanks to the dynamics of the bubbles and the escaping gas, champagne offers a playground of fascinating physics – both theoretically and experimentally. One person who has dedicated his scientific career to studying the physical chemistry of bubbles in carbonated drinks is Gérard Liger-Belair from the University of Reims Champagne-Ardenne. As well as doing science with champagne, he also consults for several champagne houses, such as Pommery, Veuve Clicquot and Moët & Chandon. For his efforts, he gets sent samples of champagne but claims not to drink any of it, saying that if he did, he 'wouldn't get any work done'.† Indeed, he doesn't have to look far for a decent supply of bubbly given that the

---

* Uncorking also results in the cork expanding by around 30 per cent.

† See physicsworld.com/a/six-secrets-of-champagne.

University of Reims Champagne-Ardenne also owns a small vineyard that produces several hundred bottles per year.

In 2019, Liger-Belair and colleagues examined what was going on in those initial few milliseconds when a champagne bottle is uncorked. And it is a lot, thanks to the images provided by high-speed cameras. The first thing that happens is a huge pressure drop in the headspace, from 5 or 7 bar to 1 bar, or atmospheric pressure, as the cork is popped.[*] The carbon dioxide gas in the headspace expands rapidly and cools, so much so that the temperature of the gas can drop to -130°C in a split second. This then results in water vapour forming together with a small amount of ethanol vapour as a visible cloud just above the opening of the bottle, which you may be able to see. The gas mixture of water vapour and carbon dioxide in the headspace escapes from the bottle at more than 400m/s – well over the speed of sound.[†] In other words, supersonic speeds.[1] This can create some wild fluid dynamics even with a humble champagne bottle. Breaking the sound barrier results in the production of shock waves, which are sharp 'discontinuities' – a sudden change – in air pressure and temperature that move as a wave. How these high-pressure waves interact with the flow (out of the bottle in this case) can then result in a visible shock wave called a 'Mach disc' that appears perpendicular to the direction of the gas flow.[‡] You can see this effect clearly in the glowing

---

[*] If you want to reduce the amount of champagne escaping the bottle when uncorked, chilled is the way to go.

[†] The speed of sound being about 340m/s.

[‡] Named after the Czech physicist Ernst Mach, who first described these types of waves in the late 1870s.

disks in the exhaust streams of supersonic jets and rockets, for example. Liger-Belair and colleagues found, for the first time, that opening a humble bottle of champagne also causes a visible Mach disc in the gas above the bottle opening.

These findings were backed up in 2022 by theoretical simulations, which found that there is even enough pressure released in the cork headspace to create a succession of shock waves that combine to form a 'shock diamond', although this has not been seen experimentally observed yet. These are repeated patterns of Mach discs, which create diamond-like structures and are also seen in the plumes of supersonic aircraft.[2] A year later, Bernhard Scheichl from the Vienna University of Technology in Austria (remember him from the dribbling teapot of Chapter 2) and colleagues also modelled the dynamics of this Mach disc, finding that it forms just outside the bottle opening between the cork and the bottle and travels away from the bottle before then moving back towards it.[3] The audible 'pop' that you hear when you uncork a bottle is a combination of the expanding cork as it leaves the bottle as well as the shock wave generated by the supersonic jet of gas, like the sonic boom you hear when an aircraft breaks the sound barrier. 'It is still not fully understood how this characteristic "pop" is produced,' Scheichl told me. Measuring the distance of the Mach disc from the bottle also provides a way to determine the gas pressure or temperature in the champagne bottle. Yet given it takes only one millisecond for the Mach disc to disappear, you might be better off with a thermometer if you *really* want to know the temperature of your champagne.

The supersonic speed of gas goes some way to explain the explosiveness of the cork as the plume pushes it out. In 2013,

Liger-Belair and colleagues opened standard 75cl bottles of bubbly at three different temperatures: 4, 12 and 18°C. They then used high-speed infrared cameras to study the cloud of carbon dioxide that is released when the cork pops. The force behind the popping process is the force exerted by the gas expanding in the headspace on to the base of the cork stopper. They found that the amount of carbon dioxide and the cork's speed both increased with the temperature of the champagne, which makes sense given that pressure is proportional to the temperature. At 18°C the velocity of the cork was almost 55km/h – enough to shatter glass – while even at 4°C the speed was still a swift 40km/h.[4] This explosiveness goes some way to explain why in 2009 the American Academy of Ophthalmology declared that champagne cork popping is one of the most common holiday-related eye hazards. If a popping cork happens to hit your eye it can rupture the eye wall, cause retinal detachment, dislocation of the lens, and even damage to the bone structure around the eye, which could require surgery.* A direct hit to the eyeball could even lead to blindness. So, when uncorking always make sure the bottle is pointed away from people – as well as lighting.[5]

Bottle uncorked without managing to lose an eye, it's time to pour it into the glass (presumably, you don't drink from the bottle). If you want to maximise your champagne

---

* If you want to do it safely, then chill the champagne before opening, don't shake the bottle, put a towel over the top of the bottle when opening and don't aim it at anyone's face.

experience, then according to Liger-Belair the popular flute shape might not be the best idea. This type of glass, which is perhaps the most popular when it comes to serving champagne, can force the bubbles to rise rapidly to the top, causing the gases to go up your nose as you take a sip – and given that carbon dioxide is acidic it might not be a pleasant experience. A wine glass, on the other hand, is too open at the rim, which allows the bubbles and aromas to escape over a larger area. A tulip glass – shorter than a flute and curved slightly inwards at the top – happens to be ideal. It still gives the bubbles space to spread a little but doesn't cause them to escape so rapidly, allowing the aromas to be savoured.

Bubbles are formed by a process called 'nucleation', which happens when clusters of gas molecules grow into bubbles via tiny imperfections, say in the surface of the glass. The irregularities of the glass's surface are too small to act as nucleation sites for champagne[*] and instead is done through the presence of tiny cellulose fibres. These could come from the air or through wiping the glass with a towel. The carbon dioxide bubbles grow on these fibres and when they reach a certain size – about 10 to 50 micrometres in diameter – they are buoyant enough to detach and float up, meanwhile another bubble soon forms in its place.

You will see many waiters or bartenders tilt the glass 45° or so and pour the liquid down the side of the champagne flutes. But does this do anything or is it just for show? To investigate, Liger-Belair and colleagues first poured champagne straight down into a vertical glass without it touching

---

[*] Some glassmakers use lasers to create artificial bubble nucleation sites at the bottom of glasses.

the sides. In this case they found that the liquid was turbulent at the bottom of the glass, which led to air bubbles forming and the carbon dioxide escaping from the champagne. While pouring 100ml of champagne vertically into a flute created an estimated 1 million bubbles, pouring down the side, however, led to a much higher concentration of dissolved carbon dioxide in the champagne and yielded tens of thousands more bubbles than straight down the middle.[6] All of which helps to keep the taste and aromas in the liquid.

Once poured and before embarking on that first mouthful, it is worth taking the time to see what's going on inside the glass. One aspect is how remarkably stable the bubbles are, seemingly rising in a perfect straight line or chain (admittedly, this is easier to see in a flute). In 2023, researchers from the US and France measured the bubble size and composition of champagne and other sparkling wines, as well as beer and sparkling water.* They then carried out fluid-dynamics simulations and experiments by pouring liquid into a small rectangular plexiglass container.[7] To study the bubble chains and what makes them stable, the team inserted a needle at the bottom so they could pump in gas to create bubble chains with different-sized bubbles. They found that large bubbles are better at forming chains than smaller bubbles. But in champagne, the bubbles are small, so something else must be going on to keep them single file. When the researchers gradually added a surfactant to the fluid they saw that smaller bubbles were better at keeping in a line. As well as giving the champagne some of its flavour, these surfactant-like protein

---

* Or more specifically Pellegrino sparkling water, Tecate beer, Charles de Cazanove champagne and a Spanish-style brut.

molecules help to reduce the surface tension between the liquid and the gas bubbles. Indeed, the surface tension of champagne is about 30 per cent less than it is for water and with a viscosity about 50 per cent more. When a bubble rises it creates a wake – a trail of motion behind it – that can affect the bubbles that come after it. Investigating further, they found that the presence of the surfactants induces a larger amount of vorticity in the bubble's wake, which creates a so-called 'negative lift force' that keeps the bubbles aligned, making for a smooth rise to the top of the liquid. In drinks that have little surfactants, such as carbonated water, or in some beverages, such as coke and beer, the bubbles instead create a smaller amount of vorticity and generate a 'positive lift force', which means that they collide and knock each other out of the way, forming a kind of cone-shaped rise of bubbles rather than a single straight line. As well as creating taste, these surfactant molecules also allow the bubbles in champagne to rise in perfect formation. Neat.

Once those bubbles reach the top of the glass they pop, releasing those surfactant odours. But they also release sound, like a gentle hissing noise, that you might be able to hear if you put your ear close to a glass. Juliette Pierre and colleagues from Sorbonne University in Paris captured this intricate process by recording the sound produced with a microphone, as well as filming what was happening with a high-speed camera. As expected, the hissing sound occurs when the bubble ruptures on the surface with the pressure inside the bubble releasing. Yet the researchers found that when this happens, a part of the bubble remains inside the liquid and vibrates the air above it until the bubble finally pops.[8] 'This vibration can be amplified by the champagne glass, that can then resonate,' Pierre told me. 'This last aspect is the reason why the sound is more prominent when the champagne is served in a thin glass instead of a plastic goblet or a mustard glass.' For small bubbles like those in champagne, only the initial rupture can be heard by humans. As the bubble then shrinks, ultrasonic frequencies – or with a frequency of about 50,000Hz – are produced, which are too high in frequency for our ears. Yet such sounds can be 'heard' with advanced microphones or by some animals, such as dogs or mice; so, if your mutt pricks up their ears near a glass of champagne you might now know why.

The big problem with opening a bottle of champagne is how long it lasts before it goes flat and all the dissolved carbon dioxide has finally made its way out. It will take about a day or so for an opened wine to go completely flat but what about unopened bottles – do they ever go flat? Even unopened bottles lose carbon dioxide as the gas escapes from the cork or metal caps. If you get given a nice bottle of champagne of a particular vintage, perhaps as a retirement or wedding present, but don't want to drink it straight away (what's wrong with you?), how long can you keep it for before it goes flat? After all, nobody wants to save an expensive bottle for a few years then, when it comes to opening it, discover that it is flat without any bubbles. Liger-Belair and his team also tackled this problem– it seems they left no stone unturned when it comes to champagne. They measured the carbon dioxide in different champagne vintages and estimated the original amount of yeast-produced carbon dioxide. They found, rather unsurprisingly, that the amount of carbon dioxide decreased as the bottles aged, but that larger bottles retained gas better than smaller ones.

With their measurements, they calculated that a standard 75cl bottle has a shelf life of about forty years, while a 1.5l and a 3l bottle have an eighty-two and 132-year lifespan, respectively,[9] meaning you are probably safe to store it away for a rainy day, especially if it happens to be a jeroboam.*

What does all this research mean when it comes to impressing your friends the next time they come to celebrate New Year or a birthday? Well, according to the latest research, serve the champagne at about 9°C, pour it gently down the side of a tulip-shaped glass, watch the bubble chains rise to the top and then sniff the aromas before feeling the tingling sensation of those bubbles in your mouth. And, more importantly, take the time to appreciate the physics that is happening in that humble glass (and bottle). The carbon dioxide bubbles rising to the top before collapsing to release all those aerosols and flavours, is, after all, the same process that happens when ocean waves crash on the beach, bursting bubbles to release the 'aroma of the sea'.

When it comes to bubbles, the most famous beverage is champagne. But bubbles are also present in many other drinks, be it a humble can of soft drink or a nice cold beer. Beer is one of the oldest and most consumed alcoholic drinks in the world, with an estimated[10] global production in 2023 of about 2 billion hectolitres.† It is generally made with four basic ingredients: water, malted cereal grains, yeast and hops. Lager

---

* A jeroboam contains 3l of champagne, while a magnum contains 1.5l.

† A hectolitre being equal to 100l.

beer involves cool fermentation and, in a similar manner to champagne, is bottled under pressure with carbon dioxide. Yet the pressure in the headspace of a chilled bottle of lager is only about 2 bar – the pressure of a standard car tyre.

Not done with just champagne, Liger-Belair and colleagues have also studied what is going on in a freshly poured lager. The team measured the carbon dioxide content of a glass of Heineken lager (at 5 per cent alcohol by volume) at 6°C and concluded that between 200,000 and nearly 2 million bubbles are created in a gently poured glass of lager before it goes flat.[11] Interestingly, they found that beer and champagne bubbles form differently in a glass, with larger imperfections in the glass leading to the production of bubbles in beer but not so much in champagne, as we discovered earlier. Even for a beer supersaturated with carbon dioxide, bubbles will not form unless they are nucleated by hitting imperfections in the glass or a particle.

When it comes to beer, the perfect pour means creating a nice foamy head, which is produced by bubbles of gas, mostly carbon dioxide, as well as chemical components, such as wort protein, yeast and hop residue. A good foamy head involves a fair amount of chemistry as the carbon dioxide bubbles rise to the surface, releasing compounds that create and stabilise the white foam. And this foamy head is important given that it has a distinct taste and is often a sign that consumers use to judge the quality of their pint. If a bartender poured you a lager and it had no foam head, you would likely refuse it. But too much can also result in an upset customer and so bartenders will, as with champagne, tilt the glass and pour it slowly down the side. This stops the carbon dioxide bubbles from escaping easily and keeps the head of the pint at a minimum.

If you go to a stadium or concert hall, or anywhere that is serving vast quantities of beer, you may have come across bottom-filling glasses in action. This is where beer is filled rapidly from the bottom of the glass, almost like magic. Of course, it isn't. The glass has a hole in the bottom that contains a ring and a thin disc-like magnet that lies over the ring. When placed on the filling station, a nozzle lifts the thin magnet up and starts to fill the cup rapidly. When the glass is filled, the bartender simply lifts it up, and the magnet falls back into place on the ring, sealing the cup so the liquid doesn't spill out from the bottom. The clever thing is that the amount of liquid to add to the glass can be programmed, so that when the cup is full, the nozzle stops automatically. This leaves the bartender to do other things, such as take payment or have a little rest – but given such machines can crank out 40 to 50 pints in a minute there is little time for that.

But what about creating the perfect head, in this case given it is usually done by carefully tilting the glass? In 2023, researchers filmed the filling process with high-speed cameras and together with simulations examined how the beer foam forms. They found that it arises in the first fraction of a second after the inlet starts to pour beer into the glass when the flow is more turbulent. When more liquid then enters the glass, it acts to push the foam to the top without creating much more bubble action.[12] The researchers also found that different temperature and inlet pressures can affect the amount of foam produced. Higher temperatures – around 15°C – and larger tap inlet pressures – about 1.5 bar – yielded more foam. They found a ratio of 20 per cent head to 80 per cent liquid occurred for a temperature of 5°C with an inlet pressure of 1 bar. So, by tweaking the temperature and the inlet pressure of the liquid, engineers can still create that perfect beer head, even when pouring a pint within a couple of seconds.

When the foamy head forms, it encounters the air and begins to decrease, so don't admire the head for too long before delving in. Yet having a big, foamy head (on the pint, that is) does confer some advantages. This is especially so when attempting to carry a few pints back from the bar to a table while avoiding other people and staying upright on a slippery floor. You may have seen those bartenders during Munich's Oktoberfest* carrying a dozen steins full of beer with seemingly little spillage (in this case, the ratio of foamy head to liquid might be close to 20/80 or perhaps even 30/70). On the other hand, trying to walk and avoid spilling a cup full of black coffee or tea can be tricky, unless you pay close attention to it or have a lid. But it happens to be slightly easier to do with beer, unless you have had a few too many drinks and are a bit wobbly.

Why is that? In 2015, researchers in the US and France wondered if the foam stops beer from sloshing around when being carried. They created an experiment to compare the sloshing of liquids with and without a foamy top.[13] This was done by placing the liquids on a moving mechanical plate and using a high-speed camera to record the sloshing waves† that were induced as the drinks were jolted around. The researchers found that liquids with foam tops hardly spilled at all, and the more foam, the more that the oscillations in the liquid were reduced. The sloshing of Heineken and

---

* Which rather confusingly often starts in September.

† Not a technical term.

Guinness,* for example, was much less compared to a cup of coffee (a plain coffee, that is, not a latte or cappuccino). To their surprise, the foam happens to be very efficient at damping the waves, even if it is just a thin layer of bubbles. This efficiency, they concluded, comes from an effect called 'viscous dissipation'. As the waves form and travel in the liquid, the foam rubs against the walls of the container, and this dissipates the energy of the wave to reduce its effects. The experiments finally explained why beer is a lot easier to carry around than coffee – and it also won't scald you in the process if it overspills, but it will create sticky floors.

And when it comes to making floors sticky with beer, one of the best ways of doing so is the irritating prank called 'beer tapping'. This is when someone taps the top of your open beer bottle with the bottom of theirs, causing a foamy mess to erupt from the bottle only for the perpetrator to some-how get away foam-free. In 2014, researchers from Spain and France discovered that when the top of a bottle is struck from above, it causes compression waves to travel down the liquid in the glass. These shock waves then rebound off the bottom of the glass as a low-pressure 'rarefaction compression' wave, then this travels upwards to the open surface of the liquid.[14] This backtracking wave just happens, however, to oscillate

---

* Another interesting thing about Guinness is that when the beer is settling, the bubbles appear to be sinking rather than rising. In 2013, physicists showed why: it is all to do with the shape of the glass. If a glass is narrower at the bottom than at the top, the flow is directed downwards near the wall of the glass and upwards in the centre, so only sinking bubbles will be seen. See Benilov, E.S., Cummins, C.P., and Lee, W.T., 'Why do Bubbles in Guinness Sink?', *American Journal of Physics*, vol. 81 (2013): 88–91.

the carbon dioxide bubbles causing them to expand rapidly. This then results in a violent implosion of the so-called 'mother' bubbles into millions of very small 'daughter' bubbles – the same bubbles that can be formed from shaking a bottle rapidly before opening it. These daughter bubbles then expand rapidly because they have a greatly increased surface area and this allows the carbon dioxide to diffuse into the bubbles faster. The bubbles quickly become buoyant, turning much of the liquid into a foam that rushes upwards, causing a foam volcano. And it only takes about half a second for all this to occur – way too quick for you to do anything about it, except perhaps put the bottle to your mouth to not lose the liquid.

Perhaps the bigger question is why the top bottle gets away with it; why doesn't a similar refraction compression wave cause a foamy mess in the perpetrator's bottle? That was a question that Johann Ostmeyer from the Helmholtz Institute for Radiation and Nuclear Physics in Bonn, Germany, pondered when a friend told him about the 2014 findings. 'I had a few pub conversations, during which I realised that I didn't understand what prevents the upper bottle from foaming over,' Ostmeyer told me. He says the reason why it doesn't is all about the shock wave. While the shock wave in the low bottle creates a lower-pressure wave in the liquid, which acts to drive the growth and subsequent implosion of carbon dioxide bubbles, the shock wave in the above bottle, however, generates a high-pressure wave that instead compresses the bubbles. In this case, they don't reach a point where they can collapse to form a foam of 'daughter' bubbles that then expand. 'Once the bubbles in the bottom bottle have collapsed into many fragments, the process is irreversible and there is nothing you can do about it,' Ostmeyer says. In other words, the prankster can leave the scene dry, unless that is you cover the bottle with your finger and aim the spray of expanding foamy liquid at them as they walk away.[15] Naughty.

Beer not only comes in glass bottles but is also sold and drunk from aluminium cans. If the prankster is still at large and you think they may have shaken the can just before you open it, or perhaps you have just dropped it on the floor, when is it safe to open the can without the liquid squirting everywhere? Some people think that gently tapping the top of the can helps. In this case, the vibration from the tapping is thought to dislodge any bubbles that have formed on the wall of the container, causing them to rise to the top of the liquid and possibly out of it into the headspace of the can before opening. But no one had studied whether this technique had any effect until 2019 when researchers in Denmark decided to carry out a rigorous analysis of the approach. The researchers obtained more than 1,000 standard 330ml cans of pilsner-style beer from Carlsberg breweries (who apparently were not involved in the study and had no vested interest in its outcome) and then somehow weren't tempted to drink all the cans before doing the study. That's because they are serious scientists and have a job to do.

They randomly assigned the cans into four groups: unshaken/untapped; unshaken/tapped; shaken/untapped; and shaken/tapped.[16] Tapping consisted of the researchers knocking the side of the can three times with a single finger, while shaking involved jolting a can for two minutes. They opened so many cans during their investigation that they had to resort to a knife to do so, as they damaged their fingers and nails. After recording the mass of each can before and after opening, they found, as expected, that shaken cans lost more fluid compared to unshaken cans – 3.5g, on average, compared to just 0.5g. Yet for both shaken and unshaken cans there was no statistically significant difference in the amount of liquid lost when tapped or not. The researchers think that the flicking is not enough to drive the bubbles to the headspace, possibly due to the flick being absorbed by the aluminium or the liquid itself. No amount of tapping the can will help those bubbles calm down, unfortunately. 'The only apparent remedy to avoid liquid loss is to wait for bubbles to settle before opening the can,' they write. Boo.

And finally in our tour of beer physics, if you have been to Argentina, you may have seen bartenders add a few peanuts when serving a beer. As a peanut is denser than the liquid, it sinks to the bottom of the glass. But after a few moments the legume will float to the surface, where it will remain for a while before sinking again, with the process repeating again and again. In 2023, researchers in Germany examined the so-called 'beer-dancing peanut' effect by painstakingly dropping whole, shelled and roasted peanuts into a litre of lager-style beer and recording what happens with a camera. They found that when the peanut was dropped into the liquid, bubbles began to attach to its surface, and they continued to do so as the peanut sank and rested on the bottom of the glass. It is thought this happens because it is more 'energetically favourable' for the bubbles to form on the peanut than, say, in the liquid or on the wall of the glass, but this process is still not fully understood.

When hundreds of tiny bubbles are attached to the peanut, it becomes buoyant and then floats to the surface.

One might think that when the peanut reaches the top of the fluid, the bubbles simply burst, but the researchers found that it is a bit more complex and instead the peanut rotates, which causes some, but not all, of the bubbles to burst (the effect isn't seen in a broken peanut, which can't easily rotate on the top of the fluid). When the bubbles burst, the buoyancy is lost, and the nut sinks back down for the process to start again.[17]

The researchers found this process repeats itself for a remarkable 150 minutes before the peanut finally comes to rest at the bottom due to the beer becoming flat. I am known to nurse my drink, but even I couldn't wait that long to replicate the test, so I will take the researchers' word for it.

Cheers!

11

FUN AND GAMES

THE DANISH PHYSICIST NIELS Bohr was a giant of twentieth-century physics. One of his most significant contributions to our understanding of the world was the model of the atomic nucleus, which now bears his name. It features an atom – containing protons and neutrons – as a central nucleus with electrons in circular orbits around it. This indicates that electrons can only occupy certain energy levels in the nucleus and are able to jump from one energy level to another by emitting/absorbing radiation. This standard diagram, or 'Bohr model', is still seen in countless physics (and chemistry) textbooks,* and for the theory Bohr bagged the 1922 Nobel Prize in Physics 'for his services in the investigation of the structure of atoms and of the radiation emanating from them'.

---

* The Bohr model was later superseded by quantum mechanical models, which describe the position of an electron in a probabilistic manner via a mathematical expression called a wavefunction.

Bohr was the leading theoretical physicist in his country and had a position created for him at the University of Copenhagen – head of the institute for theoretical physics. During the Second World War, when Denmark was occupied by Germany, Bohr fled to Sweden and eventually to the UK, where he became part of the British mission to the Manhattan Project – the US effort to build an atomic bomb that we now know so much more about thanks to the Hollywood blockbuster film *Oppenheimer*. Following the end of the war, Bohr returned to his native Denmark, becoming a staunch activist for peace, where he called on countries to freely share scientific information. He remained in the country until his death in 1962 at the age of 77.

When Bohr wasn't thinking about the very fabric of the universe, he enjoyed watching Western movies, especially the showdown duels. During such gunslinger fights, which in the movies usually features two men squaring off, waiting for one to make the first move, Bohr noticed that the good guy often drew his gun second. Despite this apparent disadvantage, he almost always came out on top, shooting the villain before they could get anywhere near their gun. Apart from teaching the lesson that the good guy always wins, Bohr thought there could be a genuine difference between such actions. The person drawing the gun first being 'willed' into action, with the second person being 'reactive' on pure instinct to shoot back. Ever curious, Bohr wanted to investigate and ordered his students to buy some cap guns from a toy shop in Denmark. Back in the lab, Bohr then challenged his students to a gunslinger fight. And he didn't just pick on the students but also challenged his colleague and fellow physicist George Gamow.

According to legend, Bohr apparently 'won' every duel. His technique was to never draw first but wait for his adversary to do so and then react to the move on instinct. He was so good that a victim even penned a four-line ditty about him, which stated that while Bohr always drew second, he still managed to shoot everyone down and how it was 'foolish' to question Bohr's wisdom.[1] Bohr's theory for his gunslinging dominance was that the brain reacts quicker to danger than it does when it carries out a deliberate intention such as drawing a gun. Bohr postulated that the reason why the villain in the movies always lost was because when they drew their gun they were confronted with feelings of shame or guilt as they actively decided to kill another human. In contrast, the hero, in this case, acts in self-defence, freeing him of such distracting thoughts. The effect became known as Bohr's law or the gunslinger effect. But does it really stack up? After all, Bohr's dominance in toy gun duels didn't exactly come down to a life or death situation.

To find out, in 2010 researchers from the UK, New Zealand, Germany and South Korea carried out a series of experiments in which volunteers competed in pressing a series of buttons as quickly as possible. The researchers simulated 'gunfights' by sitting volunteers opposite each other and asking them to hit a sequence of buttons as soon as the other person moved (no issue of being killed in this case). Each participant sat in front of three buttons, spaced about 35cm apart, with their hand lying over the central 'home' button. Their task was to press the right button first followed by the left and finally the middle button. In these mock gunfights, the initiators took about 600 milliseconds to carry out the task.[2] They found that those who 'drew second' and reacted, however, completed the movements about 21 milliseconds quicker than those who started first. Yet it still took participants about 200 milliseconds to react to their opponents moving, so the quicker time completing the task was not enough to make up the lost time by starting later.

There was still the question whether this reactive speed advantage originates from a different 'cortical' route in the brain. In other words, whether the chemical signalling really is different for a willed compared to a reactive movement. Another explanation for the effect, however, is the 'competition hypothesis', in which the reacting participant is quicker because they need to compensate for the opponent's obvious starting lead. In 2018, researchers set out to test this and again had each participant in front of three buttons. But this time the participants were told they had to work together to complete the specific movement as quickly as possible. They found that even when it came to co-operation, rather than trying to beat each other, participants who reacted to movements still executed the movement sequence faster.[3] This disproved the 'competition hypothesis', with the researchers claiming it to be evidence that signalling in the brain is indeed different for a willed compared to a reactive movement. As with the 2010 study, the quicker time completing the whole movement is unlikely to make you quicker overall when reacting second, so maybe still draw your gun first if you want to come out on top.

Given this, why did Bohr always come out victorious in his gunslinger battles? There is one explanation, and one that doesn't involve his students just letting him win. 'Our data make it unlikely that these victories can be ascribed to the benefits associated with reaction,' the authors of the 2010 study conclude. Rather, they suggest, 'Bohr was a crack shot, in addition to being a brilliant physicist.'

In Ian Fleming's novel *You Only Live Twice*, the fictional secret agent James Bond is challenged by the head of the Japanese secret service, Tiger Tanaka, to a game of rock, paper, scissors. The game involves 'throwing' one of three objects (not literally): stone donated by a closed fist, paper by a flat hand and scissors by making a 'V' with the index and middle fingers. Players start each round by saying 'rock, paper, scissors, shoot' and on 'shoot' reveal their 'throw', with rock blunting scissors, scissors cutting paper and paper covering rock. Bond and Tanaka play three games, which each involve three rounds of rock, paper, scissors. Bond's strategy in the game with Tanaka is to pick his throws at random and not react to what the opponent chooses. In the first game, Tanaka keeps it simple and plays stone in each of the three games – losing to Bond's paper in the first game, winning against Bond's scissors in the second and then losing again to Bond's paper in the decider. In the second game, Tanaka comes out on top after a series of throwing the same object. With one game each, Bond wins the first round of the final game by throwing stone against Tanaka's scissors and then wins the second by playing paper against Tanaka's stone, to be victorious by two games to one.

Bond was always going to win, of course, but was his strategy really the best way of playing according to game theory? Given there are three hand movements – rock, paper, or scissors – you would think there is a one in three chance of winning, with everyone choosing the three actions with equal probability.* Job done. Yet thanks to human emotion and decision-making strategies, this turns out not be the case. At least that is according to physicists in China. They recruited 360 students to play the game. The students were divided into sixty groups of six players with each group playing 300 rounds. Player actions were recorded, and the winners were even given money to make sure they were trying to win. After all, handing out cash is a sure-fire way to guarantee enough students will be interested; well, that and food.

The physicists found that players at the start of the games do chose each action about a third of the time, which is what you would expect if their choices were random. On closer inspection, however, the players' strategy then consists of predictable patterns – players who won the first round tended to stick with the same action, while those who lost would usually switch their throw. This usually involved rock changing to paper, paper turning to scissors and scissors into rock. So, if you played paper first and lost, there is a good chance you would change it to scissors in the next round.[4] This showed that people stick with a winning strategy and shift a losing one, also known as 'win-stay, lose-shift'. To improve the chances of winning after losing a round, a player should throw a gesture that would beat the

---

* Known in game theory as the mixed strategy Nash equilibrium, which is named after the mathematician John Nash who came up with the Nash equilibrium in 1950; see Nash, J.F., 'Equilibrium Points in In-person Games', *Proceedings of the National Academy of Sciences*, vol. 36 (1950): 48–49.

winner's previous throw. And if you win a round, to continue the winning streak anticipate that the opponent will change their throw to beat whatever throw you just made.

And what about Bond's game against Tanaka? Well, in that third decisive game, Bond could say that his strategy was random, yet Tanaka appears to have adopted the results of the study – losing in the first round with scissors, he plays the next move in the sequence – rock – in the second. If he had stuck with scissors instead and attempted to trick Bond, he would have won against Bond's paper and maybe the outcome of the overall game would have been different. It also goes to show that Fleming was well versed in the complexities of what at first might seem like a simple game.*

When deciding who should do the washing up, rock, paper, scissors is not the only way to settle something. The coin toss is considered one of the fairest ways to choose who goes first in a game – who takes the first penalty in a shoot-out, or anything else for that matter. Some of the most famous uses of the coin toss include when, in 1845, Portland, Oregon, was a coin flip away from being named Boston, Oregon. A coin flip was also used in 1903 to determine which of the Wright brothers would attempt the first controlled and sustained flight of an aircraft.

Tossing a coin usually involves flipping the coin into the air with your thumb so that it rotates several times before landing either heads up or tails up. It supposedly has a fair, fifty-fifty chance of either landing on tails or

---

* If you fancy your chances, you could try the Rock Paper Scissors World Championships, held annually in Toronto, Canada, with a top prize of C$10,000.

heads. Or does it? In 2007, the statistician – and former professional magician – Persi Diaconis and colleagues from Stanford University in the US showed that a tossed coin does not spin perfectly and this can affect the perceived fairness of the coin flip.[5] A typical coin toss lasts about half a second and it is flipped with a speed of around 9km/h. The team used high-speed videos of a coin toss to show that the person doing the tossing usually induces a small wobble or precession in the coin as it rotates through the air. In other words, a small change in the direction of the axis of rotation through the coin's trajectory. Think of a spinning top that is spinning but then the actual top of the spinning top is doing its own precession or wobble as the top spins. The outcome of this is that the coin spends slightly more time in the air with the initial side facing up, which results in a tad higher chance of landing on the same side as it started – about a 51 per cent chance on the side facing up compared to 49 per cent facing down.[*] If there is no wobble at all, then there is no 'same-side bias'.

That was the theory, but to prove there is a small statistical difference takes a huge number of coin tosses. Yet that hasn't stopped people from having a go. Around 1900, the English statistician Karl Pearson flipped a coin a mind-boggling 24,000 times, obtaining 12,012 heads, a proportion of just 0.5005. The English mathematician John Kerrich, meanwhile, during incarceration during the Second World War, passed time by carrying out 10,000 coin tosses,

---

[*] Which is not that different from the outcome of the UK's 2016 European Union referendum in which the UK voted to leave the EU by 52 per cent to 48 per cent. Perhaps it should have been decided by a coin toss.

resulting in 5,067 heads. Despite the painstaking nature of these studies, unfortunately they never recorded whether it was heads or tails that was initially facing up for the toss, so it was not possible to use the data to test Diaconis' theory.[*] In 2023, František Bartoš, from the University of Amsterdam, and forty-seven other people, conducted 350,757 coin tosses in the name of science. 'We liked the idea and challenge of collecting over 250,000 tosses to test the hypothesis empirically,' Bartoš told me. Like Kerrich, many participants carried out more than 10,000 flips. 'Everyone worried about thumb pain from flipping,' says Bartoš. 'But the only issues we experienced was shoulder pain from leaning forwards when typing the results into the computer!' The analysis revealed evidence of a same-side bias, with a probability of 50.8 per cent for a coin landing on the same side as launched – roughly the same as found by Diaconis and colleagues.[6] Interestingly, this bias varied between individuals, with some having little or even no bias, and one person in particular having a whopping 60 per cent bias.[†]

To show how much of an advantage this 50.8 per cent bias could be, the researchers make a betting analogy (of course): if you bet a pound on the outcome of a coin toss (i.e., paying £1 to enter and winning either £0 or £2 depending on the

---

[*] A test conducted in 2009 by Janet Larwood and Priscilla Ku did account for this, but the results of 40,000 coin flips were inconclusive; see www.stat.berkeley.edu/~aldous/Real-World/coin_tosses.html.

[†] The reason for the difference in bias could be that some people manage to induce more wobble or a greater precession into their coin flips. Another is that those who had more variability conducted fewer tosses, so perhaps there is an effect where the more you toss a coin the less biased you become.

outcome) and repeat the bet a thousand times, by knowing the starting position of the coin toss you could earn a cool £19 in total, on average. This might not seem much, but the researchers say it is more than the £5 that a casino would expect to make against an optimal strategy for six-deck blackjack.

Next time you find yourself deciding who should do the washing up or change the baby's nappy, give yourself a very slight advantage and go for the side that is facing up.

As the evening draws in, it's time to get out a board game. One of the most popular is Monopoly, whether playing the original, which can be traced back to the early 1900s, or the countless special editions that have been created since, such as The Hobbit, Pokémon Kanto and even M&Ms.* Regular players of the game will likely have their favourite squares to buy, willing the dice to roll the correct number so they land on it and can purchase it. Others, meanwhile, might have the strategy of buying everything they land on before they run out of money and hope that it turns out okay in the end. Monopoly is based on both chance and skill. Getting around the board mostly relies on throwing two dice, with a single turn ranging from moving two (one and another one) to twelve spaces (six and a six). But given there are two dice means there are different ways a certain number can be reached. For example, seven can be made up six ways – 1+6,

---

* It is hard to know exactly how many versions of Monopoly there are but estimates are in the thousands, see monopoly.fandom.com/wiki/List_of_Monopoly_Games_(Board).

2+5, 3+4, 4+3, 5+2 and 6+1 – which makes it the most likely to happen, with a probability of 16.7 per cent. This means that the most likely square you will hit from the start is Chance (the next most probable is rolling a six or an eight – both with a probability of 13.9 per cent – landing on either The Angel Islington or Pentonville Road, respectively, for the London edition of the game).

Going around the board once based purely on the most probable outcome results in landing on Chance; Northumberland Road; Strand; Water Works; and, finally, Liverpool Street Station.

What makes Monopoly more complicated than just pure dice probabilities is that you don't just traverse the board in this manner. If you happen to roll three doubles, on the third double you are sent to Jail. The probability of throwing a double is $6/36 = 1/6$, so for three throws it is $1/6 \times 1/6 \times 1/6 = 1/216$, or 0.46 per cent. Over fifty turns, however, the probability of it happening increases to about 23 per cent, so there is a good chance someone at some point in the game will roll three doubles.* Other moves around the board without throwing the dice include landing on Chance or Community Chest and being sent to a different part of the board, especially Jail or Go. Yet this aspect makes landing on

---

* Once you land in Jail what is the probability of leaving it by rolling the dice (rather than using a 'get out of jail card')? To leave you need to roll a double and have three separate attempts at doing so. The probability to roll the first double is 1/6, the probability of not rolling a double on the first time but the second is $5/6 \times 1/6 = 5/36$. The probability of not rolling a double on the first or second but on the third is $5/6 \times 5/6 \times 1/6 = 25/216$. The probability of rolling doubles is then $1/6 + 5/36 + 25/216 = 36/216 + 36/216 + 25/216 = 91/216$ or 42.1%. So, you have a roughly even chance of getting out of Jail by rolling the dice.

some squares more favourable than others and it turns out that Jail is the most visited space on the board.*

Once out of Jail, a player is most likely to roll six, seven or eight, landing on Bow Street or Marlborough Street (orange squares), which could be good ones to buy based on probability outcomes. As could the red squares of Strand, Fleet Street and Trafalgar Square, which would then be most likely if someone had rolled a seven when emerging from Jail. Given the rules of Monopoly, some have simulated moving around a board thousands of times to compute all the probabilities.† It turns out that orange and red squares are the most landed on together with the railway stations (mostly due to Chance/Community Chest cards taking you there), which tend to be an afterthought for most players. The least landed on squares are the maroon colours of Old Kent Road and Whitechapel Road as well as the dark blue of Mayfair and Park Lane – even though this particular pair are often favoured as they have the highest rent charges, and so if an opponent lands on one of those and you happen to have a hotel, it is often game over for them.

Another key question is what to do with houses and hotels once a monopoly of a certain colour has been purchased. Most people buy as many houses as possible to get a hotel as soon as they can to make sure they can generate the highest

---

* The strategic thinking around landing in Jail is to get out as soon as possible (i.e., by paying the ₳50 fee) early in the game to give you the opportunity to buy property, while later in the game it is best to stay inside to avoid landing on opponents' property.

† For probabilities, see this analysis from Bill Butler, www.durangobill. com/Monopoly.html.

amount of income if/when an opponent lands on the square. But is this the best return on investment given that it takes a lot of money to buy all those houses/hotels in the first place? Given the probabilities of landing on certain squares, it turns out that the quickest way to get a return on investment is to *only* buy three houses. If you own the orange properties and have three houses, then according to probabilities it will take about nine to ten opponent rolls for you to recoup the total cost. The railway stations are equally good, taking about thirteen of your opponents' rolls to recoup the investment. Compare that with the maroon properties of Old Kent Road and Whitechapel Road, where it would take a whooping twenty-nine to thirty opponent rolls to recoup the cost.*

From a probability viewpoint, the best way of playing Monopoly is to buy the railway stations as soon as you can to generate a steady investment stream and use that to buy three houses at once on the orange properties. Further property/house investment should then be carried out in the half of the board between the Jail and Go square, i.e., red and yellow properties. And always remember that your opponent is most likely to roll a seven, so you can change your strategy based on this. Of course, this statistical analysis doesn't guarantee winning every time, after all the beauty of Monopoly is in the skill as well as the chance. But abiding by these rules gives you the best chance of winning, although always bear in mind the remarks by the US author Mark Twain that there are three kinds of lies: lies, damned lies and statistics.

---

* For a summary of the probabilities, see www.tkcs-collins.com/truman/monopoly/monopoly.shtml.

Rather than battling on the Monopoly board, which often results in family arguments and usually ruins Christmas Day, you might prefer instead to spend your evening with something more calming – a simple jigsaw, perhaps. The jigsaw box usually states the number of pieces and perhaps the finished area of the Jigsaw once complete. A 2,000-piece jigsaw, for example, could take up 98 × 68cm when complete. But we have all done that frustrating thing of tipping out the contents of the box only to find that the table we are building the jigsaw on is not big enough even to accommodate all the pieces laid out, never mind being able to build the picture at the same time. This results in putting some pieces on the table and the rest elsewhere, such as in the box or even on the floor, which makes searching for that tricky piece even more taxing.

Thankfully, physicists have come to the rescue to solve this problem so you can calculate how much table space you will need to spread out all the pieces. Madeleine Bonsma-Fisher from the University of Toronto and her husband, Kent, found themselves at home during the COVID-19 pandemic with a baby and passed the time doing jigsaws. 'I like to be able to see all the pieces at once when I'm doing a puzzle, so dumping out the box and turning all the pieces over in a single layer is our customary first step in puzzle-solving,' Madeleine told me. 'Yet our dining room table is not that big, so we often feel a bit pressed for space when solving a puzzle.' They assumed that each puzzle piece is a square but then modelled the squares spread out on a table as if each were inside a circle (ever heard the phrase 'squaring the circle'?). They then analysed how these circles could be packed on to a two-dimensional area (think lying a load of non-overlapping two pence coins on a table). This forms a hexagonal packing arrangement: if you put circles all around each other and then draw a line from their centres to each neighbour it makes a hexagon.

They found that in this hexagonal packing arrangement the area of the unassembled pieces is simply √3 (or 1.73) multiplied by the area of the completed puzzle.[7] 'We were surprised by the result – I naively assumed that it would depend on the number of puzzle pieces – my intuition was that more pieces would mean a bigger unassembled area,' says Madeleine, who enjoys solving 1,000-piece puzzles in particular. 'But it turns out that the impact of having more pieces is exactly cancelled out by those pieces being smaller!' That was the theory, but does it correspond to reality? The duo completed nine different puzzles of varying sizes and number of pieces and then painstakingly measured their unassembled (by roughly laying all the pieces near each other) and assembled areas. They found that the √3 rule held regardless of the number of pieces. An unassembled 1,008-piece jigsaw had an area of 5,538 sq cm, while the completed puzzle was 50.2 × 69.0cm with an area of 3,464 sq cm, giving a ratio of 1.6. If you ever want to roughly work out how big the area your table needs to be to clearly see all the jigsaw pieces for your puzzle, simply calculate the area of the completed puzzle and then times it by 1.73. Alternatively, you could just spill the contents out on the table and hope for the best.

When you have finished the jigsaw or your game of Monopoly and you have been soundly beaten by an

8-year-old – yet again – it's time to tidy it all away and put the box in the cupboard (and the 8-year-old certainly won't be doing that part). But even this seemingly mundane aspect of putting away a board game has some surprising physics. Closing the lid on your favourite box can sometimes take a while as it slides down the base to close – you might try pushing it down to release a sound like passing wind, which an 8-year-old would certainly find funny. This so-called 'telescoping' cardboard box is where the lid barely overlaps with the base; such boxes are cheap to make, which is why they are widely used. In 2018, Kaare Jensen was tidying boxes at his lab at the Technical University of Denmark when he noticed something strange. When he put the lids on the boxes, they took different times to close and the lids fell into place in different ways; some did so slanted, while others were roughly level as they closed.*

Looking deeper into the issue, Jensen found that while the economic and environmental aspects of telescoping boxes had been well studied, the physics never had. They began to carry out experiments on commercially available boxes, including board games, as well as 3D printed models.[8] They found that a lid either closed while moving at roughly the same speed throughout or slowed down as the lid was closer to being fully closed. To investigate the fluid dynamics behind the action, the team derived a theory for the flow of a thin film of air in the gap between the lid and base. The speed of closing depends on the 'flow resistance' of the gap. As the box closes, the resistance gradually increases because

---

* For a video taken by the researchers showing different lids closing, see youtu.be/uC80c91wxyM.

the overlapping region of the lid and base gets longer, which is why lids tend to slow down as they close. They then compared this to experiments, finding that the fastest way for the box lid to close is surprisingly not based on a conventional straight lid-base configuration but for the lid to have a slight angle – just a few degrees – relative to the vertical base. It's unclear if this design will ever be used given that manufacturers are probably not too concerned by how quickly their boxes close. 'I thought we would get calls from Apple and Amazon the day after the paper was published,' Jensen told me. 'But that hasn't happened yet.' If the design ever does hit the shelves, we can at least thank the researchers for, er, thinking outside the box.

Playing a physical board game feels somewhat old school now given the onset of tablets and devices, with more and more games going online and behind screens, for good or ill. The onset of powerful algorithms and artificial intelligence is also allowing computers to beat the world's best players at many games. In game theory, perfect play is possible when a game is solved; in other words, a strategy that leads to the best possible outcome regardless of what the opponent does and even if mistakes have already been made by (both) players. There are several games that have been 'solved', such as Connect Four and noughts and crosses. A small number of games, such as draughts, have been so-called 'weakly solved', which means that an algorithm exists that can produce a win for one player, or draw for both, against any possible moves by the opponent. For some games, like chess, it is thought that they may never be solved, given their complexity. Despite this, computer programs can still play the game well by taking solutions of the game that will allow the program to play perfectly after some point. After all, in 1997 the chess grandmaster Garry Kasparov famously lost to IBM's Deep Blue, while in 2016 Go[*] 'nine-dan' Lee Sedol lost to Deep Mind's AlphaGo 4-1.[†] It was quite a feat that Lee managed to even win one game; indeed, at the time he was the only player to have ever won a match against AlphaGo.

Online chess has resulted in a huge treasure trove of data that is perfect for mining to analyse human performance. In 2009, researchers in Japan and Germany examined more than a million games from the ScidBase database,[‡] finding that the most popular opening move for white was pawn to e4, which happened in 45 per cent of games, while pawn to d4 was second most popular at 35 per cent and then third at 9 per cent was knight to f3. The most popular opening sequence of moves was the King's Indian Defence (white pawn to d4, black knight to f6 as a response, then white pawn to c4 and black pawn either to e6 or g6).[9] Another popular opening is the Sicilian Defence – white moves pawn to e4 followed by black to pawn c5 (possibly made famous

---

[*] Go is a two-player board game consisting of playing pieces called stones – one set white and the other set black. The aim is to place the pieces on the board to surround more territory than the opponent. The game is popular in Asian countries.

[†] The achievement was awarded a runner-up spot in *Science* magazine's 2016 Breakthrough of the Year award.

[‡] Available here: scid.sourceforge.net.

by the Netflix series *The Queen's Gambit*, in which the main character, Beth Harman, and others play it).

In 2022, researchers from Austria took this analysis further by looking at the difference between amateur and professional players. They analysed 123 million games played by almost 1 million players on an open-source internet chess server, lichess.org, finding that beginners tend to start with a larger array of opening first moves (perhaps because they don't know what they are doing), whereas professionals and more advanced players start their games with almost the same move (if playing white, which makes the first move).[10] Yet expert players display a broader response repertoire than beginners, showing more of an ability to surprise opponents. As a player's career progresses, they tend to specialise in few openings, rather than having a richer range of moves at the beginning. Interestingly, the researchers found that players go through streaks – both hot and cold. Players would often go on a run of wins, but would also suffer from a series of defeats, often for longer than hot steaks were observed, which the researchers say could be down to a lack of confidence, focus or 'mind fitness'.

Chess and draughts are examples of 'perfect information' games, in which everything is known about the game in progress, in other words each player can see all the pieces on the board.* Some games, however, are 'imperfect information' games in which information is hidden from certain players.

---

* There is some debate in the academic community as to whether games with chance, such as Monopoly and backgammon, can be considered perfect information. Some say they can't be because the outcome of the chance events is unknown before they occur.

One of the best examples of this is poker, in which players do not know the cards their opponents hold until the end of the game when all is revealed. Poker involves each player having their own private cards and taking turns making bets on their hands, calling other players' bets, or folding if they think they don't have a strong enough hand. The Hungarian American polymath John von Neumann* preferred poker over games like chess. He considered poker to be more like real life, in other words full of risks. Von Neumann's interest in poker was that it was not based on probability alone but other factors such as bluffing, which is possible because there are factors in the game that are hidden. It is possible to make a mistake in poker but still come out on top; likewise it is possible to 'play perfectly', but still lose the entire game. 'Real life is not like [chess],' he stated in a conversation in London with Polish British mathematician Jacob Bronowski, who was a keen chess player. 'Real life consists of bluffing, of little tactics of deception, of asking yourself what is the other man going to think I mean to do.' Despite von Neumann apparently only playing poker occasionally – and not very well – the game was the driving force behind his development of modern game theory.

The most popular version of poker today is Texas hold 'em, in which each player is given two cards face down. They peek at their 'hand' (meaning the two cards, not their

---

* Born in Budapest in 1903, von Neumann was a prodigy, publishing mathematical papers when aged 18 and in his mid-twenties was publishing a paper every month. He made contributions to numerous fields, including economics, computer science, biology, mathematics and physics, and during the Second World War worked on the Manhattan project. He died in 1957.

actual hand) and then an initial round of betting occurs. Three cards are then dealt by the dealer face up (the flop) that each player can see. Another round of betting occurs and then there are another two final rounds, where a card is added to the flop (called the 'turn' and the 'river', respectively). The players who are left (those who haven't 'folded' their hand) then reveal their two cards and the winner is the one with the best five-card poker hand using any combination of their two cards and the five community cards. In Texas hold 'em (as well as standard five-card poker), the order from most common to least common is straight (sequence of five cards in increasing value not of the same suit), flush (five cards of the same suit, not in sequential order), full house (combination of three of a kind – three cards with the same value – and a pair – two cards with the same value).*

Even with just two players, a simple fifty-two-card game can result in enormous complexity. When Texas hold 'em is played by only two players and with fixed betting sizes, it is known as heads-up limit Texas hold 'em. Despite this simplification, there are still remarkably 10 possible situations a player can find themselves in.[14] Despite such complexity, poker has still been affected by the rise of the machines. In 2015, Michael Johanson and colleagues from the University of Alberta claimed to have 'weakly solved' the heads-up version so that the algorithm would always win in the long run. It might lose some games,

---

* A list of probabilities for different types of poker hands is available at en.wikipedia.org/w/index.php?title=Poker%20 probability&oldid=1153025783.

but over many hands it would either break even or come out ahead.[11] The computer algorithm could even consider bluffing strategies where a player places big bets even with a weak hand. In each game, the program would carry out a particular strategy but after finishing the game, it would then compare it to other possible strategies and then update its approach based on the comparison. Doing this repeatedly eventually leads to what might be considered a perfect strategy. The result is that if a human spent seventy years playing 200 hands of poker per hour for twelve hours a day against the program, they wouldn't be able to tell the difference between an actual perfect strategy and the one performed by the computer.*

Taking the no-limits version, where betting is not a fixed size, expands the 10^{14} situations to a mind-boggling 10^{160} decision points. In 2017, researchers had a go at this version and came up with a program called DeepStack.† In the heads-up limit example, the algorithm computed and stored its strategy by playing a smaller game to find a solution before then 'mapping' the strategy on to the original game. DeepStack instead considers each situation as it arises during play using machine learning to give the program what might be called 'intuition'. The algorithm was put to the test against eleven professional poker players at heads-up, no-limit Texas hold 'em, who were asked to complete a

---

* If you think you are a dab hand at the heads-up limit version of poker, you can try competing against the algorithm, called Cepheus, at poker.srv.ualberta.ca. Good luck.

† See www.deepstack.ai.

3,000-game match between 7 November and 12 December 2016. Cash incentives were given to the top three performers of C\$5,000, C\$2,500, and C\$1,250. The algorithm still managed to beat every player, becoming the first computer program to outplay human professionals.[12] Those were still two-player variants of the game, so in 2019 researchers announced a 'superhuman' artificial intelligence for multiplayer poker. The program, dubbed Pluribus, learned how to play six-player, no-limit Texas hold 'em by playing against five copies of itself. When it then played five professional poker players, the computer performed significantly better than humans over the course of 10,000 hands of poker that were played over twelve days.[13]

The march of the machines continues and a year later researchers created a program called AlphaZero, which was able to master Chess and Go using reinforcement learning from self-play. Only given the rules of the game and with no embedded human strategies, it started using random play against itself, completing more games in a few hours than ever recorded in human history.[14] It was soon able to beat current state-of-the-art Chess and Go programs 'convincingly'.[15] Against the 'Stockfish' chess program, for example, it won twenty-eight games, drew seventy-two and lost none. While AlphaZero could be considered a master of perfect games, developing a general artificial intelligence algorithm that could master both perfect and imperfect games has proved tricky. That is until 2023 when researchers working at Google Deepmind created the 'Student of Games'.[16] It uses self-play learning, game theory and other techniques to learn games as it plays. It can be used to master perfect information games such as Chess and Go while also outcompeting the best available programs for imperfect information games such a Texas hold 'em poker and Scotland Yard.*

And it's not just through the process of self-play that artificial intelligence can make inroads. In 2022, an artificial intelligence algorithm named CICERO mastered the online board game Diplomacy. What was remarkable about this was that Diplomacy involves negotiation, co-operation and competition between multiple players. The algorithm was capable of conversing as well as analysing some of the intentions of its human partners in the game. Playing anonymously against humans in forty speed games in an online Diplomacy league, CICERO scored more than double the average score of human players and was in the top 10 per cent of participants who played more than one game.[17] CICERO even 'passed' as a human player against eighty-two players, with the researchers seeing no in-game messages to suggest that the human players believed they were playing with/against an algorithm.

All this might make you want to hide away and just consider playing a nice game of Monopoly or doing a jigsaw with the family, but there is little point raging against the machine when you next lose against an algorithm. When it comes to their prospective dominance, the term 'game-changer' has never been more apt.

---

* This game involves a team of detectives co-operating to track down a player (controlling a criminal, called Mr X) as they move around a board using a (finite) mix of taxi, bus and underground tickets.

THE KEY TAKEAWAY

IT'S REACHED THAT POINT in the evening when it's time to make dinner. Yet the effort of thinking what to have, sorting out the ingredients and then spending thirty minutes or so preparing it might not appeal. If it happens to be Friday night or the weekend (or any other day for that matter), then the chance to put your feet up and order a takeaway might be too tempting – and you wouldn't be alone in having such thoughts. In the UK, revenue from food delivery was almost $7 billion in 2022, with the average person spending £650 every year on their favourites.[1] In China, meanwhile, which is the largest market worldwide for food delivery, the revenue was a staggering $42.5 billion, with Meituan and Ele.me dominating food delivery in the country.[2]

Part of those huge numbers is likely because ordering food from the comfort of your own home is now so easy to do. When I was growing up, ordering a takeaway involved selecting items from a leaflet posted through the door, phoning the number to put the order through, being told it would be 'about forty-five minutes' and then searching for the cash to pay for it. With apps such as Deliveroo, Just Eat and DoorDash (which is the most popular food delivery app in the US with a 50 per cent market share) you don't even have to speak to anyone – this might be considered an introvert's dream. All it takes is a few finger swipes and you can let the app do all the hard work. They even provide minute-by-minute updates as the order is made with an estimated delivery time, and you can even track the driver/rider on a map as they get closer to your door. And thanks to Apple Pay,* there's no need to rummage through the drawers for money. As well as the ease of ordering, apps have also increased the variety on offer, with restaurants and fast-food chains now providing delivery services – an aspect that accelerated following the COVID-19 pandemic as traditional restaurants turned to deliveries to maintain business (who could forget the novelty of ordering a Sunday roast takeaway in a box complete with gravy in a polystyrene cup?). And it is not just restaurants, but also supermarkets that are getting in on the act, offering groceries delivered to your door in minutes.

As with dating or chess games, the advent of food delivery apps and the internet has resulted in huge amounts of data that can be mined for interesting trends. In 2018, scientists in the UK analysed Google searches for food-related queries, such as 'Just Eat', 'pizza delivery' or 'Chinese delivery', in Australia, Canada, India, the UK and US over a five-year period from 2011 to 2016. Despite the difference in cultures, people in each country mostly wanted a takeaway between

---

* Other payment services are available.

6 p.m. and 7 p.m. (local time), with clear spikes in internet searches for 'Just Eat' in the UK and 'Panda Express' in both the US and Canada during those times.[3] Searching for pizza delivery was also highest between 6–7 p.m. but it had another clear spike around 2 a.m. – likely from young people, perhaps university students, as they come back from an evening out and urgently need something to soak up all that alcohol.

Given that such clear trends were seen across so many different countries, the researchers concluded that the instinct to kick up your feet and have an evening off with a takeaway is 'biologically motivated'. Next time you feel peckish you could just open your favourite takeaway app and swipe and tap away, or you could resist the biological urge and instead try to make your favourite takeaway, not only impressing your friends or family in the process but also saving some money too (maybe). And thankfully, there are some physics-based tricks to help you prepare your favourite cuisine.

During excavations in 2023 of the Regio IX area of Pompeii's archaeological park close to Naples, archaeologists unearthed a still-life fresco that was dated to about 2,000 years ago. The fresco was found on a wall that is thought to have been the hallway of a home that had a bakery. The painting shows a round focaccia bread with toppings alongside various fruits on a silver tray.[4] Given that it was discovered in the birthplace of Margherita pizza, there was initially much excitement that perhaps pizza was older than we first thought. Yet this bread is likely to have been a precursor to pizza, as it lacked two staple ingredients that now make pizza what it is today: pineapple ... sorry, I mean tomato and mozzarella. These two elements are thought to have first been incorporated into pizza in the 1800s. Before that time flatbreads were most likely topped with garlic, cheese and salt. The history of pizza is much contested and there are many differing theories about its emergence in Italy. Yet in the late nineteenth century, Italian immigrants opened pizzerias across New York and, as they say, the rest is history.

Pizza is cheap (relatively), easy to make and can be customised for different tastes via toppings. It is not only one of the most popular takeaways but is also the go-to when you have people around in an evening or a children's play date to cater for. After all, is there a child who doesn't like a pizza of some sort (and especially with pineapple)? The magic of pizza lies in the dough, made from four simple ingredients – flour,* salt, yeast and water.† The pizza dough is best made a day in advance. Mix all the ingredients, knead well and leave it to prove for twenty-four hours or even more if you can. This part involves the yeast producing bubbles via a biochemical process that causes the dough to rise and the crust to become light and airy when baked. The dough can then be cut into ball-shaped portions to prove further and be ready for baking or stored in the refrigerator or freezer for later use. Once the pizza dough is stretched out to a disc, it's time to put some tomato sauce and toppings

---

* Some say an Italian Tipo 00 flour is ideal for Italian-style pizza bases.

† Although some recipes also include oil, such as Roman pizza, but a classic Neapolitan dough only contains four ingredients and is generally thicker than the super-thin Roman pizza.

on before getting it in the oven. As every Italian knows, the best pizzerias are those that have wood-fired ovens and you should never venture into a restaurant with an electric one. As the wood burns in a corner of the oven, the heat radiates uniformly through the curved walls and brick floor of the oven to ensure an even bake. Generating an even temperature throughout can take many hours, with pizzerias starting up their ovens early in the morning to get them ready for lunch.

The toppings on a pizza are mostly cooked or boiled via heat irradiation from the fire, and the pizza dough mostly cooks through conduction from the brick surface. For a wood-burning oven with a fire-brick bottom, a temperature of about 330°C inside the oven can cook a Roman thin-base pizza in about two minutes. Yet when pizzerias are busy with customers, the oven's temperature is increased to 390°C so that it takes about eighty seconds to cook the pizza – increasing the productivity of the oven by about 40 per cent. The downside of this speed bake is that the resulting pie can be overdone or a bit burnt on top and the toppings slightly undercooked. Physicist Andrey Varlamov, research director of the Institute for Superconductivity and Innovative Materials in Rome, teamed up with fellow physicist Andreas Glatz and food anthropologist Sergio Grasso to investigate what makes a good Roman pizzeria. 'We spent a lot of time speaking with pizza chefs to try and understand the best conditions to cook pizza,' Varlamov told me. 'Being theorists, we wanted to predict the experimental observations in the pizza oven.' The trio examined the theoretical physics of cooking a pizza by modelling the heat transfer in an oven – conduction from the bottom of the oven to the pizza as well as from the heat radiation from the fuel.[*] The heat-transfer equations showed that for an oven temperature of 330°C, the temperature between the brick and the pizza bottom is about 208°C (while it is 240°C for a pizza oven temperature of 390°C).[5] Assuming that the pizza dough before being cooked is about 20°C, they found that the cooking time for the perfect Roman pizza at 330°C is about 125 seconds – almost exactly the two minutes that is seen in a traditional pizzeria.[†]

Of course, not everyone has a brick oven in their garden, so what if you wanted to replicate this with an electric oven that contains a steel surface that the pizza is cooked on? The problem here is that steel conducts heat much better than brick and so if you tried to cook the pizza in two minutes, it would mean heating the interface between the pizza and oven to 300°C, which would end up incinerating the pizza – and nobody likes a carbon pizza (although there was once was a fad for charcoal pizza dough, which didn't really take off, presumably because the black appearance wasn't very appetising). According to the calculations, the best possible outcome in an electric oven requires reducing the interface temperature to around 230°C and taking 50 per cent longer to cook the pizza to make sure that the base and the toppings are cooked.

---

[*] They ignored the effects of convention, which the researchers said should be small compared to conduction and radiation.

[†] Neapolitan pizza is cooked at a higher temperature of about 460°C. This is because pizzerias in Naples often do not use fire bricks in the bottom of the oven but rather a volcanic clay produced by the nearby Vesuvius volcano. This material has a lower thermal conductivity than fire bricks, which means the pizza is not burnt.

There has also been a plethora of small pizza ovens on the market that use wood pellets or gas as a fuel together with a ceramic 'pizza stone' base that is usually around a centimetre or so thick. In this case, it is important to make sure that the ceramic pizza stone is well up to temperature to allow the pizza base to cook through (perhaps more than the manufacturer states), otherwise you risk ending up with a burnt crust on top and an undercooked base, which isn't appealing to those guests you are trying to impress. And when it comes to toppings, don't cram too many water-rich toppings, such as aubergines, otherwise they won't cook in time. Indeed, chefs in a pizzeria deal with water-rich toppings by inserting the wood or aluminium spade under the pizza and lifting it from the baking surface for another twenty seconds or so to make sure the toppings are cooked. And finally, give the pizza a lot of attention. However, whatever you do, it won't quite reach the real thing. Varlamov says that while there are many imitations, it's hard to imitate the dry heat and smell of wood in a traditional fire-brick oven.

Pizza cooked, it's time to cut it up so everyone can get a piece. Usually that involves dividing a pizza by making a series of straight cuts through the middle into four quarters and then dividing those up so there are eight equal pieces. But what if you have a couple of annoying friends, or, alternatively, annoying children who insist that they don't want to have any part of the crust? Well, you could just get new friends, but getting new children is much more difficult, so what can you do instead to stop crust wastage? Cutting a disk (or any shape for that matter) in different ways, known as tiling, is a well-researched area of mathematics and every year someone finds a new way to tile something with or without repeating patterns. In 2015, mathematicians in the UK turned their attention to various disc tilings, one of which happens to be the perfect solution to satisfy crust-phobes. They formulated a range of so-called 'monohedral disk tilings' to show the numerous ways you can split a circle into two sets of identical pieces with at least one piece not touching the centre of the circle.[6] The tricky thing here is making sure everyone gets slices that are roughly of equal size. One possibility they found results in twelve pieces: six extending from the middle (with no crust) and the other six containing quite a lot of crust (see figure below). While this results in some people having to chomp on a lot of crust, if they complain just provide a nice garlic dip with it.

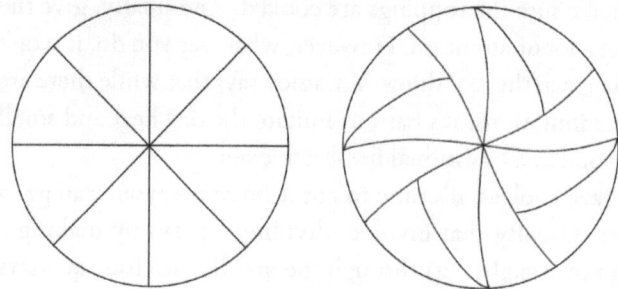

How to cut a pizza the usual way (left), and if you have some people, or children, who are 'allergic' to crust (right).

Let's presume for a moment that the pizza has been cut in the usual manner so that the pizza slice is what is known as a circular sector. What is the best way to hold it so that a floppy (but cooked!) base doesn't cause all the toppings and cheese to slide off the non-crust end and on to the floor? The secret to overcoming this toppings disaster is thanks to the mathematics of curved surfaces that was introduced by the eighteenth-century German physicist and mathematician Carl Friedrich Gauss, who is perhaps best known for his work in magnetism.* In 1828, Gauss came up with *Theorema Egregium* (Latin for 'remarkable theorem'), which concerned curves and surfaces and established the notion of curvature. A major consequence of the theory is that surfaces of different curvature, such as a sphere (which we say has a positive Gaussian curvature), cannot be transformed on to a flat (zero curvature) surface without distortion. This is the exact problem that goes into designing geographical maps, when the sphere of the globe is transposed on to a flat surface such as a screen or sheet of flat paper for a poster. You can try it yourself: take a tomato or orange and cut it in half and then scoop out the innards so that just the skin is left. Now if you try to flatten the tomato half so that it is perfectly flat on the table, it isn't possible without tearing the skin in some way. The same goes for taking a flat piece of wrapping paper and trying to wrap a football perfectly. It's not possible without overlaps or tearing the paper itself.†

Back to pizza. When you hold the pizza by the crust, the other end flops down due to gravity. After all, the pizza is thin and thin sheets are easy to bend, so gravity takes over. Without going into all the details, for a flat surface like pizza, Gauss' theory says that if one direction is curved the

---

* The unit of measurement of magnetic induction or magnetic flux density is the Gauss (G).

† Shapes that can transform into each other without stretching, such as a flat piece of paper being rolled into a cylinder, are called 'isometric', whereas a ball and a flat sheet are not isometric.

other will be flat.* In the case of the droopy pizza, the length of the pizza slice is now curved (downwards) while the wide portion stays flat. Not helpful. What we need instead is to make the length flat. To achieve this involves curving the width of the pizza by folding it. In this case, the only way to make the length droop would be to deform the pizza by stretching it, and gravity alone can't do this. You can try it yourself – and there is no need to light up the pizza oven. Just take a sheet of paper and hold it at one end as you might do a pizza slice and see that the other end will flop down. Now while holding it in the same place, this time curve the width of the paper, so it looks like a skateboarder's half-pipe – the length becomes magically flat with no droop in sight, all thanks to Gauss' 'remarkable theorem'.

In 2022, the Italian theoretical physicist and Nobel Prize winner Giorgio Parisi caused a furore in Italy after suggesting how to cook pasta. The physicist, who works at the Sapienza University of Rome, won the Nobel Prize in Physics a year earlier for his work in the theory of complex systems, in particular how disorder and fluctuations can together give rise to something that is predictable. Parisi's 'crime' was to suggest turning down the gas to finish cooking the pasta. Sacrilege! He shared someone else's comment on Facebook that recommended Italians add pasta to a pan of boiling water, bring it to the boil, wait two minutes, then put a lid on and turn off the gas and cook the pasta for one

---

* For a nice video about this, see youtu.be/HGl3_92KW7I.

more minute than prescribed. The post claimed this would save some eight minutes of energy consumption. Parisi added a subtle change of his own, however, claiming that domestic cooks should tighten the lid and instead of turning the gas off, put it at its lowest setting. 'A lot of heat is lost through evaporation,' he noted, adding that a tight lid is 'essential'. Despite the focus of the advice being to help Italians with a cost-of-living crisis caused in part by Russia's invasion of Ukraine in early 2022, his observations immediately landed him in, er, hot water. After all, Italians are incredibly protective of their cuisine and chefs soon called him out, adding that turning down the heat would be a 'disaster' and result in rubbery pasta.[7] 'Let's leave cooking to chefs while physicists do experiments in their lab,' noted Italian chef Luigi Pomata.[8]

Like pizza, pasta is a very popular cuisine, equally among children, who seem to go through a phase of eating nothing else other than macaroni, spaghetti, fusilli or whatever other form of pasta you put in front of them.[*] In 2023, researchers at Nottingham Trent University examined the energy requirements of cooking pasta, which usually involves putting 100g of pasta into 1 litre of boiling water for about ten minutes. The team found that about a third of the energy came from bringing the water to the boil, but a huge 60 per cent was in boiling the pasta for ten minutes. Carrying out Parisi's technique, however, halves the cooking costs and it

---

[*] Physicists love pasta so much that in 2014 they even created their own version. Davide Michieletto and Matthew Turner from the University of Warwick invented 'anelloni', which is a fresh pasta that has a large ring shape. You can find details and a recipe for how to make it at physicsworld.com/a/a-taste-for-anelloni.

would be even more effective for induction hobs that take some minutes to cool down once turned off.[9] The researchers also discovered that soaking the pasta in cold water for two hours to rehydrate it before cooking saved the same amount of energy as turning off the gas. But there is more. Do you really need a litre of water for 100g of pasta? Turns out not. When using half the amount of water* – and thus saving money on the cost of heating it – they found that the quality of the cooked pasta was unchanged.† I can only presume the researchers have been banned from entering Italy.

Pasta is composed of starch granules and as it cooks it grows in volume and softens – known as hygroscopic swelling – to become edible. This begins from the outside and as the water gradually penetrates into the strand, the proteins and starches break down and the pasta softens. For *al dente* pasta, for example, the outer layers are soft with a slightly hard inner region. Spaghetti is a particularly interesting type of pasta for physicists given that the initially rigid rods become elastic when cooked, which results in different type of behaviour. If you stand a handful of spaghetti strands in hot water in a pan, the straight rods will remain vertical but as cooking commences will first start to sag and then once settled on the pot's bottom will curl up to form a 'u' shape – a process known as sagging, settling and curling.

---

* Yet the team found that reducing the water further to a third did have a negative impact on the quality of the cooked pasta.

† A similar conclusion was reached in a 2018 study; see Cimini, A., Cibello, M., and Moresi, M., 'Reducing the Cooking Water-to-dried Pasta Ratio and Environmental Impact of Pasta Cooking', *Journal of the Science of Food and Agriculture*, vol. 99, no. 3 (2018): 1258–1266.

Gravity alone is not enough to describe this effect – if a straight elastic rod is bent by gravity when placed on a flat surface it will return to its initial shape rather than staying curled up. In 2020, Nathaniel Goldberg and Oliver O'Reilly of the University of California, Berkeley, soaked single spaghetti strands in room-temperature water for several hours and took images every fifteen seconds so they could study the process slowly. 'I've always been fascinated by the manner that spaghetti absorbs water and elongates, swells, and softens in the process,' O'Reilly told me. The team then developed a model using a theory that was first developed by the Swiss mathematician Leonhard Euler in 1744. This so-called 'rod theory' was modified to capture the spaghetti changing length, diameter, density and elastic modulus* as it soaked and contacted in the pan.[10] They found that the model could emulate the three different shapes as it did so – sagging, settling and curling – and could replicate the changes from being rigid and brittle to having 'viscoelastic' properties. In other words, how the material's fluid and elastic properties change over time and cause it to bend in such a way that it doesn't break. They also found that the strand becomes permanently curved after being soaked – a phenomenon that also occurs in growing plant stems. But the researchers don't recommend 'cooking' pasta for hours at room temperature given that the bonds between starch molecules in the pasta only start to break down at about

---

* Also known as Young's modulus. It is a measure of a material's resistance to elastic deformation or stiffness. A high-modulus material stretches very little when pulled, while a low-modulus material can stretch easily. It is measured in the unit of pressure: the Pascal or Pa.

50 to 60°C. 'Pasta must be cooked in warm water in order for a process known as starch gelatinisation to occur,' says O'Reilly. 'This process adds flavour and texture to the pasta – a fact that I didn't fully appreciate until after I tasted pasta that was soaked in cold water for two hours.' Urgh.

During the COVID-19 pandemic when everyone began working from home, mechanical engineer Jonghyun Hwang, who was based at the time at the University of Illinois at Urbana-Champaign, along with Jonghyun Ha and Sameh Tawfick, searched for interesting things to study around their homes. 'When the lab closed, we researchers had nothing better to do than looking for interesting scientific phenomena physically closer to home,' Hwang told me. They settled on spaghetti. 'Pasta noodles, to our eyes, were an interesting choice of material, as they are long and slender and go soft when cooked in water – and were something that we could find at home.' Hwang and colleagues began to study the swelling and softening as pasta cooks[11] by submerging spaghetti strands in hot water and cooking them for thirty minutes. Every few minutes they would take out the stands and measure certain parameters such as the elastic modulus, a measure of the spaghetti's 'softness'.

They found that the softness during cooking does not increase linearly with time, but rather exponentially, going from an elastic modulus of 10^9 Pascal (in other words doesn't stretch) to 10^5 Pascal* (able to stretch) – a difference of four orders of magnitude. The pasta softened more than a thousand times between six and twelve minutes into cooking

---

* In comparison, a rubber band has an elastic modulus of around 10^7 Pascals.

alone, with a dramatic drop near nine to ten minutes – the time that most pasta manufacturers recommend as being *al dente*. 'That's probably why cooking perfect pasta is so hard,' adds Hwang. 'Aiming for the right timing to remove the noodle from the pot requires some luck.' The researchers also looked at the role salt plays during cooking, which is added to improve the texture and taste of cooked pasta. They found that salt affects the chemical and mechanical properties of the pasta to delay softening. If you cook a noodle for nine minutes in salted water it is likely to be about twenty times stiffer than noodles cooked in distilled water, for example. The researchers think the role that salt plays when cooking is the reason why manufacturers put a range of cooking times on the packet: cooking time all depends on how much salt we add – or not – to the pan.

It might only take a couple of minutes to go from under to overcooked, so the critical question is how do you know when it is done perfectly? The classic way is to throw a strand across the room at the wall and see if it sticks (what are you, a toddler?). The team in Illinois came up with another way to determine that classic *al dente* texture. The researchers placed noodles very close together – about 5mm apart – and then cooked the strands in a beaker of boiling water. Every so often, they lifted the strands out of the water and looked at how they stuck together. They found that adhesive forces – or the degree to which the pasta sticks together thanks to surface tension – increases with cooking time. But by examining how many of the strands of spaghetti are *not* stuck together, a chef can work out if their pasta is cooked to perfection. When cooked *al dente*, which took about ten minutes in their case, they found that the free length, i.e., the length of pasta *not* stuck together, was about 20mm. Any less and it risks being overcooked, while any more and it might not be cooked at all. 'All you have to do is to leave a ruler in your kitchen,' adds Hwang.

Enough with the practical advice, what about the biggest mystery of all: why does dry spaghetti always break into several pieces if you try to snap it and not just two pieces? This tricky question even flummoxed the US theoretical physicist and Nobel Prize winner Richard Feynman, who, along with his friend William Hillis, spent hours with broken spaghetti strands all over the kitchen. Despite their effort, they failed to come up with a theoretical explanation for the effect.* It wasn't solved until 2005 when two physicists from France took high-speed images of individual strands of spaghetti as they broke and came up with a model based on a thin elastic rod. When the pasta strand is bent, elastic energy is stored in the rod and when broken it releases as a 'bending' wave down the pasta rod. It was thought that these waves would help the two halves of the spaghetti to relax, but the researchers found the opposite.[12] The bending waves instead increase the curvature of the strand, stressing it further and triggering an avalanche of new breakages, which can create more waves and more fragments.†

In 2018, Jörn Dunkel, Ronald Heisser and colleagues from the Massachusetts Institute of Technology went to

---

* It's not just spaghetti. At the London 2012 Summer Olympics, pole vaulter Lazaro Borges's pole snapped in three as he attempted a jump; see youtu.be/VrHiK1aHWL0.

† The pair were awarded the 2006 Ig Nobel Prize in Physics for their work.

extreme lengths to find a way to snap a strand of spaghetti in only two pieces. Heisser, who was a student at the time, spent a month with a friend in their dorm room bending hundreds of spaghetti strands, just like Feynman and Hillis. 'We caused a huge mess breaking spaghetti everywhere and didn't clean it up much,' Heisser told me. 'Every now and then we popped to the grocery store to get another box.' Through trial and error by stressing the spaghetti in different ways, he came up with the idea that twisting the spaghetti strands as they were bent might produce just two pieces. They then spent $2,000 (not their own money, they were students, after all) building a device to test the idea. The resulting contraption held the spaghetti in place at each end and then was able to precisely twist the strand at one end. The other end then moved slowly to the other so that the strand bent before snapping.[13]

For the actual experiments in the lab, Dunkel, Heisser and their colleague Vishal Patil used three spaghetti types: Barilla No. 1, 3 and 5, which all have slightly different diameters. They then filmed the process as it happened with a high-speed camera, and not any typical high-speed camera, but one that could take 1 million frames per second. Even then it almost missed the fracture in the spaghetti as it snapped. 'Even with this camera, how the crack propagates through the spaghetti could only be seen in five frames,' Dunkel told me. They found that to make a piece of spaghetti break into only two pieces required twisting the strand by at least 150° (with 360° being a full rotation) before bending it. The key aspect is that, when twisted, the strand bends less before breaking as compared to not twisting the strand. This means that the bending wave that occurs when the spaghetti breaks is not as strong as in the untwisted version and so doesn't go on to make further breaks. In other words, the torsional effects of twisting overcome the effects of the bending. However, experiments involving bending untwisted strands at different speeds revealed varying number of fragments produced after breaking, the fastest resulting in up to seven fragments and the slowest often producing just two.

Given that Feynman died in 1988, we will never know what he might have thought about the work. 'I hope he would have enjoyed it,' adds Dunkel. In any case, it seems that Hillis did. He saw the published research and got in touch with the researchers, thanking them for their 'delightful contribution [to] this puzzle'. '[Feynman] and I played around with twisting, but our experimental methods were not as good as yours,' Hillis wrote in an e-mail to Dunkel.

If pizza or pasta is not your thing (really?) physics still has you covered. The second most popular takeaway choice worldwide is Chinese (although it is actually the top in the UK and the US).[14] When cooking a stir-fry, there is nothing more satisfying than hearing the sizzle of food as it hits the blisteringly hot oil in a frying pan or wok. But how do you know when the oil is hot enough to add ingredients? A common technique, especially in Asia, is for chefs to moisten bamboo chopsticks with water (hopefully not saliva) and dip them into the oil in the pan. Other techniques involve flicking a bit of water into the oil or throwing in a small piece of batter. When moistened bamboo chopsticks enter the oil, small bubbles (yep, more bubbles) form, perhaps due to the evaporation of the moisture or the sudden expansion of trapped gas. These bubbles then expand, release from the chopsticks and rise to the surface. By watching the bubbles that form on the chopsticks, as well as listening to the associated crackling sound emitted by the bubbles, chefs can judge whether the temperature of the oil is hot enough to start cooking. Professional chefs can be so good at it that they can use their chopsticks to deduce the temperature of the oil in the pan to an accuracy of within 10 per cent.

In 2022, Akihito Kiyama, who at the time was working at Utah State University in the US, and colleagues examined the physics underlying this clever test. They heated a Pyrex beaker containing 125ml of canola oil to about 200°C, and used sensitive microphones and high-speed cameras to observe what was going on when a thin wet piece of paper (representing a thin piece of food), moistened chopsticks or water droplets were dropped in.[15] When the team inserted a moistened chopstick into oil that was a 150°C, they found that the chopstick was covered in a layer of small bubbles. At 180°C, meanwhile, larger bubbles began to form, especially at the tip of the chopstick.* The team think that the hotter the oil – and so the bigger the temperature difference between the chopstick and the oil – the quicker the bubbles expand. At 210°C, large bubbles formed all over the chopstick and at this temperature the researchers noted that the

---

* It is thought that bubble formation is caused by many factors, including the surface texture of the chopsticks. Larger bubbles are observed at the tips of the chopsticks, which contain small holes that can act as nucleation sites. The sides of the chopsticks, meanwhile, tend to produce small bubbles – something that is also seen using metal chopsticks.

sound emitted from the spitting oil began to change. Indeed, it is this temperature that is most favourable for cooking and goes some way to show that the chopstick test really does work. 'By both watching bubbles and hearing the sound associated with it, chefs can judge if the temperature of the oil is ready for frying,' Kiyama told me. He adds that while using different chopsticks could lead to different outcomes, the chefs use their keen sense of hearing to get around the problem. 'The perception of visual and audio information could be interrelated, and this deserves further study,' adds Kiyama. 'But chefs are perhaps unknowingly adjusting these feedback mechanisms by experience in their own kitchens.'

Once you have determined the oil is hot enough, it's time to add some ingredients. Fried rice, or scattered golden rice, is a dish that dates back 1,500 years to the Sui Dynasty (AD 581–618). The stir-fry technique involves tossing the items in a wok that has ideally been heated up to a scorching 1,200°C, with fried rice being one of the most challenging variants. At such high temperatures, under the so-called Maillard reaction, named after the nineteenth-century French chemist Louis Camillie Maillard, food quickly browns as amino acids from proteins react with sugars, imparting a desirable flavour – *wok hei* or 'breath of the wok'. This only takes a minute or so, but if left for too long the sugars can easily caramelise and burn. David Hu (of wet dog-shake fame in Chapter 4) and Hungtang Ko from Georgia Institute of Technology became interested in the physics of cooking and noticed that while much work had been carried out on western cooking – pizza, pasta, as we have just seen – little attention had been paid to Chinese cuisine.

At the time, Ko was on military service in Taiwan and couldn't perform experiments in a lab, but he had a 'really good' fried rice restaurant nearby. Between 2018 and 2019, he filmed five chefs in Taiwan and China, who each had some twenty years of experience, as they took just a couple of minutes to make fried rice.[16] This covered about two minutes of preheating, adding ingredients and then stirring, with the researchers creating a film of 276 cycles of wok tosses. At such high temperatures, and with such a heavy wok, the key is to keep the rice moving to avoid it burning. Chefs don't lift the heavy iron wok off the heat, which can weigh some 1.5kg, but instead slide it around in such a way that makes sure the fried rice is not just a burnt mess. 'If you are wanting to cook thousands of bowls of fried rice a day, you want to do it in a way that you don't get injured,' Hu told me.

The team discovered that wok experts used two alternating movements: a seesawing up and down tilt that catapults the grains into the air, almost like a ski-jump for the fried rice, and a towards-and-away sliding motion of the wok to make sure the rice grains are not only caught and do not fall out of the wok but also slide to the end of the wok, where it can be flung into the air again. Through simulations, it turns out that chefs optimise this process, having a good combination of a seesawing and back and forth motion. This makes sure that the rice is cooking in the wok as well as constantly leaving the wok in a way that maximises the distance the rice travels in the air so it cools a little and ends up being perfectly browned and not burnt. '[Ko] didn't think all this would end up being real research,' says Hu. 'But it turns out it is the work that he most famous for.'

The researchers found that the chefs tossed the rice 2.7 times per second, with the motion of the wok resulting in the rice having the maximum 'air time'. In the blink of an eye, the chef is flinging rice into the air, catching it, mixing it and then flinging it into the air again. Yet that constant flipping means that some chefs develop significant injuries. A study in 2011 surveyed more than 750 wok-using chefs in Taiwan, finding that 85 per cent had suffered at least one injury within the previous year. Some 63 per cent reported shoulder strain, with a similar percentage for neck pain.[17] Hu thinks that robots could help. While some exist, such as one at Changi Airport in Singapore, these machines usually focus on mixing the ingredients while frying and operate at much lower temperatures than that needed to make good fried rice. Hu and colleagues even attempted to build a robot cooking arm using the findings from their study but found that it broke down frequently. It seems that getting a machine to replicate what the human arm can achieve is a tough ask. 'The human arm is an amazing thing,' adds Hu. 'For the power that a human arm can deliver, it turns out chefs are doing the best they can.'

Dinner is over and there is one last thing to do: wash it all up. Boo. If you are lucky to have a dishwasher, it is not only good for the environment but saves a lot of time, too. The time spent loading and unloading a dishwasher is about nine minutes per day, on average, compared to the sixty minutes or so that would be needed to clean the day's dishes by hand, especially if you have managed to burn the fried rice in the wok. Dishwashers also have an average water and energy consumption of about 13l and about 1 Kilowatt hour (kWhr), respectively, compared to a massive 50l and almost 2kWhr for handwashing.

Engineer Raúl Pérez-Mohedano from the University of Birmingham was so convinced that there was more scope for improvement that he carried out his PhD in the 'cleaning principle in automatic dishwashers'.* Mohedano produced a theoretical model, backed up by experiments, for how water is distributed in a dishwasher. To reveal what was going on inside, he used a technique called Positron Emission Particle Tracking (PEPT). This involves introducing a radioactive particle, known as a tracer, into the system and imaging it as it moves inside to produce three-dimensional dynamics of the particle's movement. In the case of a dishwasher, a radioactive particle† was included in the water system and by placing a dishwasher between PEPT cameras (that are each about the size of a dishwasher) they could then watch as the particle moves about in a typical cleaning cycle.

The scientists loaded a Whirlpool dishwasher with twelve dinner plates, twenty-four dessert plates, twelve

---

* For those interested, it can be found at pure-oai.bham.ac.uk/ws/ files/18089860/Perez_Mohedano_et_al_Positron_emission_particle_ tracking_Chemical_Engineering_Journal_2014.pdf.

† An isotope of fluorine, ^{18}F, was used as a tracer. The isotope has a half-life of about 110 minutes and undergoes beta decay, releasing a positron. When the positron collides with an electron, the two particles annihilate, producing two gamma rays with an energy of 511 kilo electron volts (KeV) that are emitted back to back, 180° apart. The PEPT detector can determine the position of the positron to within a few millimetres.

teacups, twelve glasses and twelve bowls. Their loading method involved dinner plates and dessert plates placed in the lower basket, while small crockery items were put in the upper basket. This is a common way to pack a dishwasher unless you are a psychopath – large and heavy items, such as saucepans, bowls and dinner plates, being placed in the lower basket while delicate glassware and small items, such as cups, are placed in the medium/upper basket.

Mohedano and colleagues also used egg yolk to spoil the plates to see how well they were cleaned. Yolk can be stubborn to remove, as it hardens when dried. It was previously thought that the jets of water were enough to get rid of the dirt on a plate, effectively shearing it off the crockery. But the researchers found instead that while the tracer particle shot out of the spinning spray arms rapidly and in a straight line, when it hit an object, it slid down slowly. They deduced that shear stress is probably not the full story when it comes to clean plates and the chemical dissolution of dirt through a detergent is likely a much bigger factor (who knew!).[18]

One critical finding of all this research that you *can* use in your day-to-day life is that dishes are best washed when separated by at least 20mm. Any less and they risk not being cleaned, so try not to pack the dishwasher too much. But you probably didn't need a PEPT to deduce that. And one final tip: according to science you don't need to pre-rinse your plates before you put them into the dishwasher. Domestic argument averted!

13

PHYSICS ON THE SCREEN

THE END OF THE day, and alas this book, is fast approaching. Before it's time to hit the sack and get some much-needed kip, how about chilling on the sofa and watching a film or perhaps curling up with a good book (such as this, I hear you say)? Movies are an important part of who we are: they mirror our beliefs, struggles and the complexities of being human to provide entertainment, escapism and a window on to complex social issues. Cinema offers a wide range of genres catering to different tastes and interests. If you have company for the evening then there will probably be discussions, or arguments, over what film to stream (anyone still have DVDs or VHSs?), often involving statements such as: 'I don't fancy watching an action film' or 'I don't want to watch a romcom'. If you have kids, then there is only one winner: it really must be *Mario Movie* for the fourth time this week.

Everyone has their favourite film, or several favourites, that they watch on a semi-regular basis even though they know every scene inside out, perhaps even a significant proportion of the movie's script. I watch *Home Alone* every Christmas, but not at any other time of the year, and can recall most of the dialogue in *Jurassic Park* – a film I first saw when I was 11 and have rewatched tens if not a hundred times since. The film industry is about bringing stories to life on the big screen, with the adaptation of novels or biographies being commonplace, thanks to their compelling narrative and well thought out characters and plots (I am embarrassed to admit, however, that I haven't read Michael Crichton's *Jurassic Park*).

Despite a wide range of genres in both book and film, there is a surprising homogeneity when it comes to narrative storytelling. That is according to US fiction author Kurt Vonnegut, who outlined six main emotional arcs to describe the different types of stories.* They are: 'rags to riches', in which a subject of the film has an emotional rise; 'riches to rags' being the opposite; 'man in a hole' shows the subject's fall and subsequent rise; 'Icarus' being the opposite – a rise followed by a fall; 'Cinderella' highlights a rise-fall-rise pattern; and 'Oedipus' has a fall-rise-fall pattern. In 2016, researchers at the University of Vermont used computer algorithms based on natural language processing to analyse some 1,327 novels in Project Gutenberg's digital fiction collection.[1] The study involved examining 10,000-word chunks of text at a time and comparing the language used as the story developed. After crunching all the data, they confirmed what Vonnegut deduced – our most beloved

---

* You can watch a short lecture by Vonnegut about the 'simple shape of stories' at youtu.be/oP3c1h8v2ZQ.

works are in fact based on well-trodden narratives – and that storytelling is both an art and a science!

Given that novels fall into these six main narrative arcs, it is little surprise that movie scripts do, too. After all, scriptwriters need to condense an entire novel or non-fiction book into a two-hour film without losing the essence of the story or nuance of the characters. *Shawshank Redemption*, for example, is 'rags to riches', while *Godfather* is a 'man in a hole' movie, with *Babe* being 'Cinderella' and *The Little Mermaid* 'Oedipus'. In 2020, researchers in the UK examined 6,174 full-length movie scripts to see how successful these six categories are at the box office in terms of both revenue and when it comes to bagging awards. While any emotional arc-type film can be a success, viewers prefer 'man in a hole', with those films generating $37 million on average – perhaps due to their ability to attract viewers' attention.[2] This compares to 'Cinderella' at $33 million and 'Oedipus' at $31 million. 'Rags to riches' and its opposite, meanwhile, were bottom at $29 million, on average. 'Man in a hole' and 'Oedipus' also happen to be the most downloaded e-books, showing that these emotional arcs apply well across both written and visual forms. Interestingly, when it comes to ratings as listed on the website IMDb,* however, 'man in a hole' is the lowest among viewers and film critics, while 'riches to rags' is highest among critics ('rags to riches' is the highest rated, on average, among viewers). Yet the 'man in a hole' arc tends to generate the most 'talked about' movies rather than the 'most liked'. The researchers say this shouldn't result in movie producers only making 'man in a

---

* www.imdb.com.

hole' movies, however, as all movie types can do well. What is important is having a great script and actors, as well as a sensible budget. Easy then.

When it comes to big hits at the box office, there is only one winner. As anyone with a Disney+ subscription knows, the so-called Marvel Cinematic Universe is huge. It consists of films based on the American comic books published by Marvel Comics and features all your favourite superheroes, including Spider-Man, Iron Man, Captain America and (perhaps) the Incredible Hulk. As of 2023, the series was the highest-grossing film franchise ever with a total worldwide box office revenue of $29.55 billion – more than the next three 'cinematic universes' put together: Star Wars ($10.33 billion), Harry Potter ($9.58 billion) and James Bond ($7.88 billion).[3] Of the individual films, *Avengers: Endgame*, released in 2019, is Marvel's highest-grossing movie, taking $2.8 billion in global revenue from a budget of $356 million. Kerching!

In 2020, Matthew Roughan and colleagues from the University of Adelaide in Australia used tools from information theory to examine what makes this series so successful. 'The idea of networks in narratives has been around for a while, and that intrigued me,' Roughan told me. The tricky thing for the researchers was getting accurate data on who starred in the movie. You might think this is an easy task – just look at the credits. But there are many ways that people can appear and there are no rules about which characters get a 'name' and who is credited on a movie. This includes being a main character (such as appearing in most of the action and dialogue), a bit part (having a few lines of dialogue), an extra (a smaller role, providing background in particular scenes) or a cameo (a single-appearance role, usually by a well-known person). The team used data from IMDb to determine 'named' characters in the film, such as 'Iron Man', or unnamed characters, which are usually 'indirectly named', such as 'Tony Stark's secretary'. They also determined those that had 'minor roles' but are uncredited in IMDb lists.

The researchers wanted to study how the characters in the franchise interact, but looking at IMDb data on who performed in a movie doesn't give you that. So, rather than simply examining how much dialogue a particular character has, they instead looked at their prominence based on how many 'conflicts' they have, which makes sense given that action films tend to be conflict driven. To do so, the team had to painstakingly watch all the Marvel movies and then transcribe instances of conflict between, say, a hero and a villain. The researchers took notes while watching the films, which required frequently pausing and rewinding the movie to make sure they got their assertions correct – a Herculean effort. 'I told a friend I was going to rewatch all the Marvel movies over the Christmas holidays,' Roughan told me. 'He bet a bottle of wine that I couldn't write a research paper about it. I won the bottle of wine.'

The researchers determined a 'conflict-based' population size for a movie's cast by using theories devised by ecologists when studying the interaction between species in a habitat.[4] This resulted in a conflict-based cast size of 119 characters in the Marvel Universe, compared to 248 that would be acquired by simply adding them all up as they appear in the movies. They found that *Ant-Man*, for example, has a run-time of 117 minutes and has eighty-two conflicts and 865 lines of dialogue. *Avengers: Age of Ultron*, meanwhile, has a whopping 302 conflicts and 975 lines of dialogue in the film's

141 minutes. They found that 'origin' movies – in which a character is first introduced to the audience – have much smaller cast sizes than their sequels. For example, *Iron Man 2* has a cast inflation of 14.8 per cent over *Iron Man*, while *Iron Man 3* has a cast inflation of almost 100 per cent compared to the first movie. This happens to a similar degree with all Marvel movies (the exception being *Guardians of the Galaxy Vol. 2*, which has a smaller effective cast than the first film – *Guardians of the Galaxy*). In general, the average cast size of Marvel films increases with an annual growth rate of about 5 per cent, which shows that it is not just *our* universe that is expanding. Origin films also feature far more dialogue compared to sequel films, which are more conflict driven. After all, origin films require the audience to get to know the characters and for a plot to develop, whereas sequels can get straight into the action (and conflicts) from the get-go. So while *Iron Man* has eighty-one conflicts, *Iron Man 2* has 112 and *Iron Man 3* features 121– despite the similar run time for the three movies.

Despite the movie industry generating large revenues – estimated to be about $77 billion in 2022[5] – being profit-able is another matter. Between 2000 and 2010 only a third of Hollywood movies made a profit, i.e., had a higher box office revenue than its production budget.[6] Yet the Marvel Universe defies such trends. The researchers in Australia found that most Marvel films take twice as much at the box office than they cost to make and all but one – *The Incredible Hulk**– are better than break even. In most cases, the larger

---

* *The Incredible Hulk* was not well received and the lead actors were replaced in later movies.

the cast size the more profitable the film, with sequels (more action, conflicts and characters) doing better than origin movies. The authors believe these so-called 'team-up' films do so well at the box office and in ratings because they attract fans of several different 'origin' movies (*Avengers: Infinity War* having an IMDb rating of about 8.5, compared to 7 for *Thor*, for example).

There is, however, one stand-out movie even compared to the team-up movies: *Black Panther*. Released in 2018, it made three and a half times what it cost to make and one reason for its success is not only the quality of the acting and script, but also the timing of its release. It was seen as a rallying cry for more diversity and representation in film and changed how Black women were depicted on screen. 'There are a bunch of factors that go into the relative success of each movie, and I don't want to trivialise the importance of the acting, direction and other supporting artists and technicians,' says Roughan, adding that his favourite Marvel movie is *Captain Marvel*. 'I love the fact that they stomped on the accepted wisdom that female superheroes weren't commercially viable,' he says. 'And *Captain Marvel* is a really great movie.'

Yet when it comes to diversity in film, things might not always be as they appear. *Wonder Woman*, released in 2017, became one of the highest-grossing superhero origin stories of all time, bringing in more than $822 million, and is still one of the highest-grossing films that was directed by a woman: Patty Jenkins. The film, made by Warner Bros and based on a DC Comics character, was praised for its depiction of a strong female lead character. But what do we mean by 'female lead'? In what way?* In 2018 social scientist Pete Jones from the University of Manchester used the dialogue in the film to create a network of the main characters.[7] As expected, the analysis identified two main characters, Steve Trevor (played by Chris Pine) and Diana (played by Gal Gadot), with the remainder being smaller parts in terms of dialogue. Most of the dialogue in the film is between Trevor and Diana but, when considering all the dialogue, 56 per cent of the lines are spoken by men and 43 per cent by women. Indeed, Pine has 265 lines to Diana's 222. Further analysis revealed that most of the female characters only appeared in the first half of the movie, leaving just two – Dr Maru and Etta Candy – with whom Diana can interact later in the film.[†] This, Jones states, means that Diana's 'narrative power' is through her relationship with Trevor and so he questions whether the film is truly female led.

Jones compared the results of this analysis to the Marvel origin movie *Thor*, released in 2011. Again, there are two dominant characters in the film: Thor (played by Chris Hemsworth) and Jane Foster (played by Natalie Portman). When analysing dialogue, Thor speaks fewer lines compared to other typical lead superheroes, but even then, Jane still speaks fewer lines than Thor. In eighteen of the Marvel Cinematic Universe films from *Iron Man* in 2008 to *Black*

---

* There is already the Bechdel test, invented by the cartoonist Alison Bechdel in the mid-1980s, which uses a set of criteria to evaluate how well a work of fiction represents women. This includes if the work features at least two female characters; if they talk to each other; and if the characters discuss something other than a man. See bechdeltest.com.

† *Wonder Woman* does, however, pass the Bechdel test.

Panther in 2018, women make up 28 per cent of named speaking characters but deliver just 22 per cent of all spoken lines. And even in other so-called 'female-led' films, such as *Lara Croft: Tomb Raider* (2001) or *The Hunger Games* (2012) – and their sequels – the percentage of lines spoken by women is less than or around 50 per cent.

All of which shows that despite some progress in the last few decades, when it comes to dialogue and gender representation,[*] at least according to network science, Hollywood still has some way to go.

Big productions are not just the premise of the big screen, TV series in the past decade have become ever more popular thanks to the rise of streaming services such as Netflix and Disney+. While traditional TV series may have been scheduled – and watched – a set number of times per week, now viewers can binge an entire season in a day if they wish. One of the biggest small-screen adaptations in recent years has been *Game of Thrones*, which portrays a fictional society that undergoes constant political upheaval, civil wars and a lot of violence. The series is based on George R.R. Martin's *A Song of Ice and Fire* fantasy books, of which there are

---

[*] And not just gender but also ethnic diversity. A study in 2024 found that white actors feature more frequently and more prominently on posters for US-produced films than people of colour. The faces of white actors were 25 per cent larger on average and located closer to the centre of the poster. See Kagan, D., Levy, M., Fire, M., et al., 'Ethnic Representation Analysis of Commercial Movie Posters', *Humanities and Social Sciences Communications*, vol. 11 (2024): 180.

currently five: *A Game of Thrones* (released in 1996), *A Clash of Kings* (1998), *A Storm of Swords* (2000), *A Feast for Crows* (2005) and *A Dance with Dragons* (2011).* The books have sold over 90 million copies and been translated into forty-seven languages – I'm not sure if this book will quite reach those heights. Part of the series' success is Martin's ability to create a universe that is rich with characters, plots and sub-plots (see also *Harry Potter* or *The Lord of the Rings*). In 2007, HBO optioned the series for TV adaptation, the first episode of which aired in 2011. It was an instant hit and resulted in eight seasons that overtook the narrative of the books. On 19 May 2019, 13.6 million people around the world tuned in to watch the last-ever episode of *Game of Thrones* as it aired on HBO, rising to 19.3 million within twenty-four hours.

One feature that kept *Game of Thrones* audiences on the edge of their seats – for both book and screen – was who would live or be killed off. This can be hard to predict given Martin's ability to shock readers, and indeed much of the discussion and hype around the final TV series was about who would meet their end – perhaps in a gruesome way. In 2020, researchers in the UK and Ireland scrupulously read each book, noting the interactions between characters to create a network among characters. Such analysis has been used for films before. In the Hollywood network, the nodes are the actors, and a link is added between them when they have jointly appeared in a film together. This has spawned the Kevin Bacon game, which consists of trying to connect

---

*   George R.R. Martin is expected to release the sixth book – *The Winds of Winter* – at some point, which he has been working on for over a decade.

a given actor to the *Footloose* star and seeing if you can manage to guess the shortest possible path of collaboration in film.* For example, Tom Cruise has a Bacon number of one because he starred in the film *A Few Good Men* with Bacon, while Anthony Hopkins has a Bacon number of two, because he was in *Audrey Rose* with Kathryn Janssen, who starred with Bacon in *Only When I Laugh* (Hopkins having never worked directly with Bacon).†

In the network the researchers created for *A Song of Ice and Fire*, characters are represented by a 'node' and the interactions between them are given as a link or edge, as well as information about when a character died. From the 343 chapters, they identified 2,007 characters, of which 1,806 interact at least once with another character. Examining the links gives an indication about the main characters in the series: the larger the nodes, the more dominant the character. Some of the biggest nodes in the network include Jon Snow, Jaime Lannister and Tyrion Lannister. The books are arranged such that each chapter contains a particular person's 'point of view', with the books featuring twenty-four 'point of view' characters in total. Most of the biggest nodes are these point-of-view characters. Yet there are also three non-point-of-view characters who have a high degree of connectivity between other characters: Robb Stark, Stannis Baratheon and Tywin Lannister.

---

* Available at oracleofbacon.org.

† There is a science-based version of the Kevin Bacon game, which is the Erdös number. It is the closest someone has worked with the Hungarian mathematician Paul Erdös, who died in 1996 and who published more than 1,500 papers with 492 co-authors.

Like the discussion around origin films, the study found that the first book in the series – *A Game of Thrones* – features a small number of characters, about twenty to thirty per chapter. In later books that number is more like thirty-five characters per chapter, and even up to eighty-nine for Chapter 16 in *A Feast for Crows*. Yet this number – around thirty – is an important one. In a social network, it is known as a stable number for a subgroup.[8] The researchers say that despite the introduction of new characters throughout the series, which keeps the overall number of characters high (despite the deaths), Martin manages a consistent social network structure that is at the top end of what readers can handle. This means that fans can keep on top of things without getting lost and overwhelmed in all those names and relationships and losing the essence of the story.

The researchers also looked at deaths as they are told in the chapters of the books (which is called 'discourse time') and compared it with the number of deaths by date ('story time'). They found that the number of deaths as they came chapter by chapter had an average of between zero to three, with the occasional massacre of seven or so. But when they looked at story time,* they found the deaths were much more clumped together, with many deaths occurring between June 299 to July 300 but relatively few from October 297 to June 298. This, the researchers say, shows that the narrative structure gives the impression that deaths can occur unpredictably, yet there are large periods of *story time* where no deaths occur. 'Portraying significant events by discourse time instead of as

---

* The opening of *A Game of Thrones* is thought to take place in the year 297, with *A Dance with Dragons* ending in the year 300.

they happen appears to maintain the reader's suspense,' they write, which basically means that you keep turning that page to find out what is going to happen next.

However, despite it being difficult to guess what comes next in the series, that hasn't stopped some people from going to some extreme lengths to see if mathematics has an answer. In 2014, statistician Richard Vale from the University of Canterbury in New Zealand used the number of chapters dedicated to each character in previous books to predict the likelihood that they will have chapters in the next two upcoming books, with no chapters meaning the character is killed off. To do this Vale used a technique called Bayesian statistics, which is a method to deduce the probability of future events. Formulated in the 1700s by the English statistician, philosopher and Presbyterian minister Reverend Thomas Bayes, Bayesian analysis is a technique that calculates the probability of an event happening by considering not only the likelihood of it taking place but also information about what previously occurred, what is known as 'prior beliefs'. These two aspects are combined to produce a 'posterior belief', and this is iterated based on new evidence until a good estimate is found. Bayesian statistics is a particularly powerful method to pick out trends when data are probabilistic, with the technique used extensively in finance, medicine and genetics. The model assumes that the longer a character has been around, the more likely they are to be killed off. One of Vale's predictions is that the key point-of-view character Jon Snow, whose fate is left as a cliffhanger in book five of the series, has a 60 per cent chance of still being alive in book six and 40 per cent in book seven.[9]

Turn on the TV in the UK and there is a good chance it will be a cookery programme. They first began as educational shows, providing easy-to-follow cooking instructions for viewers, but soon morphed into a form of entertainment, with celebrity chefs having an increasing influence on consumer food habits and choices, whether good or bad. It's fair to say we spend a lot more time watching cookery shows than we do in the kitchen. According to a YouGov poll in the UK, *The Great British Bake Off* is the most popular food and drink programme of all time, followed by *MasterChef*, *Ramsay's Kitchen Nightmares* and *Ready Steady Cook*.* It is also not just daytime or prime-time TV where you can learn how to prepare avocado on toast,† but there are entire channels devoted to food, such as the Food Network. In 2014, it was estimated that 434½ hours of cookery were shown in the UK on terrestrial and non-terrestrial television.‡

In the early 2000s, cookery shows entered the era of 'reality TV', with one of the most successful being the UK series *Come Dine with Me*. It began in 2005 and features five amateur chefs who live in the same area but don't know each other. On each day during the week (Monday to Friday), one

---

* Retrieved from yougov.co.uk/ratings/entertainment/popularity/all-time-food-drink-tv-programmes/all.

† In 2015, the UK chef and celebrity Nigella Lawson caused a stir after showing viewers how to make avocado on toast; see www.mirror.co.uk/tv/tv-news/nigella-lawsons-avocado-toast-recipe-6765060.

‡ Retrieved from www.dailymail.co.uk/tvshowbiz/article-2771553/How-fed-434-hours-TV-cookery-week-cook.html.

of them has a go at hosting a three-course dinner party for the other contestants. After the party is over, the competitors rate the host on a scale of zero (terrible) to ten (perfect) on the evening they provided, which is done in private in a taxi on the way home. Whoever racks up the highest score at the end of the week wins £1,000 and bragging rights.

The success of the show has led to several spin-offs (celebrity and 'professional' versions) and has gone international in more than thirty countries, such as in Germany, *Das Perfekte Dinner*, or the US, *Dinner Takes All*. In 2020, two physicists from the University of Hamburg – Peter Blum and Marc Wenskat – analysed 2,268 episodes of *Das Perfekte Dinner* that were aired between 2006 and 2019, as well as 540 episodes of the German celebrity spin-off, for any trends. They found success in the show had a surprising similarity with a problem that was first introduced in 1960 in a column in the magazine *Scientific American* by the US popular science writer Martin Gardner. This so-called 'secretary problem' concerns a known number of candidates who are interviewed one after the other for the job of a secretary. Each one is rated immediately after the interview, i.e., accepted or rejected, with the decision to accept or reject only based on the relative rank of the applicants who have already been interviewed.

The difficulty for the hiring manager is that they don't know the quality of the yet unseen candidates. This throws up the question of what the optimal strategy is that maximises the probability of selecting the best candidate.* The shortest known general proof of this is the 37 per cent rule.

---

* This being like the issue of parking the car that we came across in Chapter 6.

The optimal point to stop is to reject the first n/e applicants, where n is the number of people being interviewed and e is 2.71828182 ... – the base of the natural logarithm. If you have ten candidates, then you should reject the first four, and doing so means you have a 37 per cent chance of picking the best candidate. You might think, why 37 per cent? The aim is to obtain a good enough baseline – a bit like shopping around when buying a sofa, for example. The difference here is that you cannot go back to the first shop if that happens to be the best. So, within the rules of the secretary problem, establishing a baseline by rejecting a given number of candidates offers an optimal solution while not necessarily getting *the* best candidate. The idea is that you then accept the next offer that is better than the baseline (or the last if none happen to be better).

There are other solutions, one of which is called the cardinal pay-off variant, in which if there is a random distribution of outcomes between two fixed numbers, say zero and one, then the 'solution' is to reject $\sqrt{n}$ -1. While in the classic version, the objective is to try to get the best applicant, the cardinal version is more about selecting a higher-valued applicant over a lower-valued one while accepting it might not be the best. For our ten applicants, for example, this would mean rejecting the first $\sqrt{10}$ -1, or two people. In *Come Dine with Me*, for example, where the number of applicants is five (or sometimes four), this would mean rejecting $\sqrt{5}$ -1, or one contestant. In this case, rejecting means offering the first contestant a low score, or one that could easily be beaten.

When the researchers looked through the data in *Das Perfekte Dinner*, they found that contestants pretty much followed the cardinal variation.[10] Of course, there are some differences with the secretary problem and what is going on in the TV show, given that the judges are the candidates themselves (and are not simply rejected or accepted) and the grading is also subjective. While the average number of points across the week was 7.57 (slightly higher in the professional version at 8.05), when they looked at the average score for each day of the week, they found that Monday was 7.4 while Friday was 7.7. Examining the probability of winning, they found only 10 per cent of those who hosted the dinner on Monday won – a percentage that rose to around 25 per cent for Friday and Thursday. To compound things, if the first person does win, they seldom do alone and instead share the winnings with someone on another night by getting the same score. This happened in 62 per cent of cases in which someone who hosted on Monday won. On the other hand, if the last contestant wins (i.e., on a Friday), they mostly always do so alone.

It had been suggested that one tactic for the person who has hosted the evening early in the week would be to give a low score to subsequent participants, a so-called 'cooking bias'. The person who cooked on Monday would purposely give lower scores to everyone else to boost their chances of winning. Yet the researchers found that this was not the case; the average score awarded by someone prior to cooking was always lower than afterwards – perhaps they were relieved to get it over and in a better mood. For example, those who cooked third awarded 7.59 prior to cooking, but 7.77 after, while those who cooked fourth awarded an average of 7.57 before cooking and 7.76 afterwards. You might think that this is because the contestants get to know each other better as the week goes on and therefore get more sympathetic to each other's circumstances. But given that similar trends occur in the celebrity version of the show, this hypothesis can perhaps be dismissed.

As in the secretary problem, the first contestant is being used as a benchmark and no matter how they perform they will receive a score that can be bettered by another contestant. If a Monday contestant does win (alone), they have probably done an exceptional cooking/hosting job to have defied the odds. *Come Dine with Me* might be a popular show, but it is also an unfair game. Yet that can be extended to other competitions that have a sequential order of performance and with scores given as soon as the person has finished. 'The easiest way to make it fairer would be to just rate all contestants at the end of the week,' Blum told me. 'This way one could compare all performances before ranking them, making the use of the secretary problem strategy unnecessary.' Another is to use a 'decoy' who goes first but doesn't take part in the competition. While those strategies are unlikely to be used in the programme, if you ever find yourself on the show and you cook on a Monday, you might as well just open a jar of sauerkraut.

So how would Blum approach being on the show if he was ever asked? 'I would cook wild boar goulash,' says Blum. 'And, more importantly, I would ask if I could go last.'

EPILOGUE

WE HAVE REACHED THE end of the day – and this book. Along the way we have encountered how physics is integral to many activities we carry out day in, day out. Whether that is in the morning when simply making a cup of coffee or tea, getting frustrated at how a steel razor blade can blunt so easily on soft hairs, or when a sopping wet dog performs a wet-dog shake after arriving home following a quick walk in the rain. Many of the discoveries outlined in this book have been made by scientists following their curiosity as they go about their daily lives and then applying physics principles to unearth new phenomena or reveal the inner working of everyday occurrences. As we have seen, getting frustrated at combing their daughter's hair not only led to a mathematical analysis of combing but the design of a robot that one day could carry out the task, allowing carers to do other things with their time. Even something as mundane as being kept awake at night listening to the characteristic 'plink' made when a water droplet falls on to the surface of a liquid resulted in an explanation for how underwater bubbles can produce sounds that propagate in air.

As the day continues and we head out to pursue our hobbies and activities, physics also rears its head in the process of blooming flowers as we tend the garden to the intricate aerodynamics that can occur when playing our favourite ball sport and the mechanics at play when moving around town during the daily commute. In areas that you might not suspect physics has much to say, such as love and relationships, playing your favourite board game or watching a movie or TV show, a mathematical treatment can reveal certain rules. Who knew, for example, that the hit TV show *Come Dine with Me* is a great example of the so-called secretary problem in probability theory (as is an optimal strategy when it comes to parking your car along a single-street parking lot!).

As the Sun sets and the day ends, it is now time to turn off the TV or put down your book (not this one, just yet) and get some sleep to perform the same routine once again tomorrow. Hopefully, the next time you carry out some of these activities you will do so with a little more knowledge about the physical processes that underline them – and use that know-how in your daily life to beat the kids at Monopoly, make perfect *al dente* pasta or pour a glass of champagne in a way that maximises the wine's flavours.

It might be a cliché to suggest that physics is everywhere, but the topics explored in the previous pages show that our regular routines are filled with physics at every turn. Physics is not just about abstract theories or complex equations that are confined to textbooks or laboratories. It is a subject that can explain extraordinary phenomena, such as colliding black holes in the depths of deep space, as equally well as the seemingly mundane effect of liquid dribbling down the underside of a teapot's spout.

Physics not only shapes the cosmos but every aspect of our lives.

ACKNOWLEDGEMENTS

THIS BOOK WOULD NOT have been possible without the help of many people who kindly gave their time to discuss their research either through Zoom or e-mail. In no particular order: Bernhard Scheichl, Christopher Hendon, Frank Schreiber, Nafisa Begam, Sunghwan Jung, Jonathan Touboul, David Dunsten, Christofer Clemente, Johann Ostmeyer, Sidney Redner, František Bartoš, Cyril Rauch, Carlos Gershenson, Kaare Jensen, René Ledesma-Alonso, Steffen Bakker, Anurag Agarwal, David Quéré, Raymond Goldstein, Madeleine Bonsma-Fisher, Andrey Varlamov, Charles Gerba, Dmitry Solnyshkov, Matthew Roughan, Michelle Gaines, Jörn Dunkel, Ronald Heisser, Peter Blum, Oliver O'Reilly, Ardian Jusufi, Katja Söhnel, Jonghyun Hwang, Akihito Kiyama, David Hu, John Crimaldi, Cemal Cem Tasan, Jason Steffen, Pete Jones, Wenjing Lyu, Benjamin Seibold, Mostafa Nabawy, Mouad Boudina, Christian Fankhauser, Andrew Dickerson, Juliette Pierre, Liza Sopina, Gareth Tyson, Nicholas Nelson, Federico

Battiston, Gérard Liger-Belair and Richard Sear. Despite their help to make the book as scientifically accurate and error-free as possible, any mistakes that remain in the text are all mine. Apologies also to anyone who I may have missed.

I also would like to thank my agent Andrew Lownie for his support with this project as well to Amy Rigg and Rebecca Newton at The History Press. Last, but certainly not least, to my family: Claire, for letting me spend time in the evening writing this book when I perhaps should have focused my energies on planning that kitchen refit, and to Henry and Elliott for their endless imagination and curiosity, and for making every hour of every day so much more fun. Thank you also to Henry for drawing the fish (his favourite animal) for the figure in Chapter 4.

NOTES

Introduction
1 Fajzel, W., Galbraith, E.D., Berrington-Leigh, C. et al., 'The Global Human Day', *Proceedings of the National Academy of Sciences*, vol. 120, no. 25 (2023): e2219564120.

Chapter 1
1 Méndez Harper, J., McDonald, C.S., Rheingold, E.J. et al., 'Moisture-Controlled Triboelectrification During Coffee Grinding', *Matter*, vol. 7 (2023): 1–18.

2 Wadsworth, F.B., Vossen, C.E.J., Heap, M.J. et al., 'The Force Required to Operate the Plunger on a French Press', *American Journal of Physics*, vol. 89 (2021): 769–775.

3 Gianino, C., 'Experimental Analysis of the Italian Coffee Pot "Moka"', *American Journal of Physics*, vol. 75 (2007): 43–47.

4 Navarini, L., Nobile E., Pinot, F. et al., 'Experimental Investigation of Steam Pressure Coffee Extraction in a Stove-top Coffee Maker', *Applied Thermal Engineering*, vol. 29 (2009): 998–1004.

5 King, W.D., 'The Physics of a Stove-top Espresso Machine', *American Journal of Physics*, vol. 76 (2008): 558–565.

6 Acquired from sca.coffee/sca-news/25-magazine/issue-3/defining-ever-changing-espresso-25-magazine-issue-3.

7 Deegan, D.R., Bakajin, O., Dupong, T.F. et al., 'Capillary Flow as the Cause of Ring Stains from Dried Liquid Drops', *Nature*, vol. 389 (1997): 827–829.

8 Gulka, P.C., Swartz, J.D., Trantum, J.R. et al., 'Coffee Rings as Low-resource Diagnostics: Detection of the Malaria Biomarker Plasmodium falciparum Histidine-Rich Protein-II Using a Surface-Couples Ring of Ni(II)NTA Gold-Plated Polystyrene Particles', *ACS Applied Materials and Interfaces*, vol. 6, no. 9 (2014): 6257–6263.

Chapter 2

1 Henrywood, R., and Agarwal, A., 'The Aeroacoustics of a Steam Kettle', *Physics of Fluids*, vol. 25 (2013): 107101.

2 Reiner, M., 'The Teapot Effect … A Problem', *Physics Today*, vol. 9, no. 9 (1956): 16–20.

3 Scheichl, B., Bowles, R.I., and Pasias, G., 'Developed Liquid Film Passing a Smoothed and Wedge-shaped Trailing Edge: Small-scale Analysis and the "Teapot Effect" at large Reynolds Numbers', *Journal of Fluid Mechanics*, vol. 926 (2021): A25.

4 Boudina, M., Kim, J., and Kim, H-Y., 'Amplitude of Water Pouring Sound', *Physical Review Fluids*, vol. 8 (2023): L22002.

5 Bi, Xiaotian, Su, D., and Zhou, Q., 'Why do Hot and Cold Water Sound Different when Poured?', arxiv.org/abs/2403.14740.

6 Rosato, A., Stranburg, K.J., Prinz, F. et al., 'Why the Brazil Nuts Are on Top: Size Segregation of Particulate Matter by Shaking', *Physical Review Letters*, vol. 58 (1987): 1038–1040.

7 Hong, D.C., Quinn, P.V., and Luding, S., 'Reverse Brazil Nut Problem: Competition Between Percolation and Condensation', *Physical Review Letters*, vol. 86 (2001): 3423–3426.

8 Gajjar, P., Johnson, C.G., Carr, J. et al., 'Size segregation of Irregular Granular Material Captured by Time-Resolved 3D Imaging', *Scientific Reports*, vol. 11 (2021): 8352.

9 Vella, D., and Mahadevan, L., 'The "Cheerios Effect"', *American Journal of Physics*, vol. 73 (2005): 817.

10 Begam, N., Ragulskaya, A., Girelli, A. et al., 'Kinetics of Network Formation and Heterogeneous Dynamics of an Egg White Gel Revealed by Coherent X-ray Scattering', *Physical Review Letters*, vol. 126 (2021): 098001.

11 Roura, P., Fort, J., and Surina, J., 'How Long Does it take to Boil an Egg? A Simple Approach to the Energy Transfer Equation', *European Journal of Physics*, vol. 21 (2000): 95.

12 Buay, D., Foong, S.K., Klang, D. et al., 'How Long Does it Take to Boil an Egg? Revisited', *European Journal of Physics*, vol. 27 (2006): 119–131.

13 Retrieved from newton.ex.ac.uk/teaching/CDHW/egg.

14 Di Lorenzo, E., Romano. F., Ciriaco, L. et al, 'Periodic Cooking of Eggs', *Communications Engineering*, vol. 4 (2025): 5.

Chapter 3

1 Brown, S.P., and Bhushan, B., 'Durable Superoleophobic Polypropylene Surfaces', *Philosophical Transactions of the Royal Society A*, vol. 374 (2016): 20160193.

2 Phillips, S., Agarwal, A., and Jordan. P., 'The Sound Produced by a Dripping Tap is Driven by Resonant Oscillations of an Entrapped Air Bubble', *Scientific Reports*, vol. 8 (2018): 9515.

3 Roscioli, G., Taheri-Mousavi, S.M., and Tasan, C.C., 'How Hair Deforms Steel', *Science*, vol. 369 (2020): 689–694.

4 De Berker, D.A.R., André, J., and Baran, R., 'Nail Biology and Nail Science', *International Journal of Cosmetic Science*, vol. 29 (2007): 241–275.

5 Ruach, C., and Cherkaoui-Rbati, M., 'Physics of Nail Conditions: Why do Ingrown Nails Always Happen in Big Toes', *Physical Biology*, vol. 11 (2014): 066004.

6 Plumb-Reyes, T.B., Charles, N., and Mahadevan, L., 'Combing a Double Helix', *Soft Matter*, vol. 18 (2022): 2767–2775.

7 Hughes, J., Plumb-Reyes, T., Charles, N. et al., 'Detangling Hair Using Feedback-driven Robotic Brushing', IEEE 4th International Conference on Soft Robotics (2021): 487–494 doi: 10.1109/RoboSoft51838.2021.9479221.

8 Gaines, M.K., Page, I.Y., Miller, N.A. et al., 'Reimagining Hair Science: A New Approach to Classify Curly Hair Phenotypes via New Quantitative Geometric and Structural Mechanical Parameters', *Accounts of Chemical Research*, vol. 56, no. 11 (2023): 1330–1339.

9 Leon, N.H., 'Structural Aspects of Keratin Fibres', *Journal of the Society of Cosmetic Chemists*, vol. 23 (1973) 427–445.

10 Goldstein, R.E., Warren, P.B., and Ball, R.C, 'Shape of a Ponytail and the Statistical Physics of Hair Fiber Bundles', *Physical Review Letters*, vol. 108 (2024): 078101.

11 Keller, J.B., 'Ponytail Motion', *Society for Industrial and Applied Mathematics Journal of Applied Mathematics*, vol. 70, no. 7 (2010): 2667–2672.

12 Touboul, J.D., 'The Hipster Effect: When Anti-Conformists All Look the Same', *Discrete and Continuous Dynamical Systems – B*, vol. 24, no. 8 (2019): 4379–4415.

Chapter 4

1 Retrieved from www.ukpetfood.org/resource/uk-pet-food-s-annual-pet-survey-shows-cost-of-living-impact-on-pet-owners.html.

2 Ratschen, E., Shoesmith, E., Shahab, L. et al., 'Human-Animal Relationships and Interactions During the Covid-19 Lockdown Phase in the UK: Investigating Links with Mental Health and Loneliness', *PLoS ONE*, vol. 15, no. 9 (2020): e0239397.

3 Muanga, M., Byberg, L., Egenvall, A. et al., 'Dog Ownership and Survival After a Major Cardiovascular Event', *Circulation: Cardiovascular Quality and Outcomes*, vol. 12 (2019): e005342.

4 Marsa Sambola, F., Williams, J.M., Muldoon, J. et al., 'Quality of Life and Adolescents' Communication with their Significant Others (Mother, Father, and Best Friend): The Mediating Effect of Attachment to Pets', *Attachment and Human Development*, vol. 19, no. 3 (2017): 1–20.

5 Reis, P.M., Jung, S., Aristoff, J.M. et al., 'How Cats Lap: Water Uptake by *Felis Catus*', *Science*, vol. 330 (2010): 1231–1234.

6 Crompton, A.W., and Musinsky, C., 'How Dogs Lap: Ingestion and Intraoral Transport in *Canis familiaris*', *Biology Letters*, vol. 7, no. 6 (2011): 882–884.

7 Gart, S., Socha, J.J., Vlachos, P.P. et al., 'Dogs Lap Using Acceleration-Driven Open Pumping', *Proceedings of the National Academies of Sciences*, vol. 112 (2015): 15798–15802.

8 Griffin, T.M., Main, R.P., and Farley, C.T., 'Biomechanics of Quadrupedal Walking: How Do Four-Legged Animals Achieve Inverted Pendulum-Like Movements', *Journal of Experimental Biology*, vol. 207 (2004): 3545–3558.

9 Bishop, K.L., Pai, A.K., and Schmitt, D., 'Whole Body Mechanics of Stealthy Walking in Cats', *PLoS ONE*, vol. 3 (2008): e3808.

10 Haagensen, T., Gaschk, J.L., Schultz, J.T. et al., 'Exploring the Limits to Turning Performance with Size and Shape Variation in Dogs', *Journal of Experimental Biology*, vol. 225, no. 21 (2022): jeb244435.

11 Wilson, A.M., Lowe, J.C., Roskilly, K. et al., 'Locomotion Dynamics of Hunting in Wild Cheetahs', *Nature*, vol. 498 (2013): 185–198.

12 Rottier, T., Schulz, A.K., Sohnel, K. et al., 'Tail Wags the Dog is Unsupported by Biomechanical Modeling of Canidae Tails Use During Terrestrial Motion', *bioRxiv*, doi.org/10.1101/2022.12.30.522334.

13 Dickerson, A.K., Mills, Z.G., and Hu, D.L., 'Wet Mammals Shake at Tuned Frequencies to Dry', *Journal of the Royal Society Interface*, vol. 9 (2012): 3208–3218.

14 Krsmanovic, M., Ali, M., Biswas, D. et al., 'Fouling of Mammalian Hair Fibres Exposed to a Titanium Dioxide Colloidal Suspension', *Journal of the Royal Society* Interface, vol. 19 (2022): 20210904.

15 Krsmanovic, M., Ghosh, R., and Dickerson, A.K., 'Fur Flutter in Fluid Flow Fends off Foulers', *Journal of the Royal Society Interface*, vol. 20 (2023): 20230485.

16 'Photographs of a Tumbling Cat', *Nature*, vol. 51 (1894): 80–81.

17 Kane, T.R., and Scher, M.P., 'A Dynamical Explanation of the Falling Cat Phenomenon', *International Journal of Solids and Structures*, vol. 5 (1969): 663–670.

18 Zhu, Y., and Shi, F., 'Why Does the Goldfish Disappear in the Fishbowl?', *The Physics Teacher*, vol. 47 (2009): 424–426.

19 Schluessel, V., Kreuter, N., Gosemann, I.M. et al., 'Cichlids and Stingrays Can Add and Subtract "One" In the Number Space from One to Five', *Scientific Reports*, vol. 12 (2022): 3894.

Chapter 5

1 Retrieved from www.surrey.ac.uk/sites/default/files/2021-07/garden-report.pdf.

2 Retrieved from www.gardenpatch.co.uk/gardening-statistics.

3 Dumais, J., and Forterre, Y., 'Vegetable Dynamics: The Role of Water in Plant Movements', *Annual Review of Fluid Mechanics*, vol. 44 (2012): 453–478.

4 Nawkar, G.N., Legris, M., Goyal, A. et al., 'Air Channels Create a Directional Light Signal to Regulate Hypocotyl Phototropism', *Science*, vol. 382 (2023): 935–940.

5 Chelakkot, R., and Mahadevan, L., 'On the Growth and Form of Shoots', *Journal of the Royal Society Interface*, vol. 14 (2017): 20170001.

6 Liang, H., and Mahadevan, L., 'The Shape of a Long Leaf', *Proceedings of the National Academies of Sciences*, vol. 106, no. 52 (2009): 22049–22054.

7 Huang, C., Wang, Z., Quinn, D. et al., 'Differential Growth and Shape Formation in Plant Organs', *Proceedings of the National Academies of Sciences*, vol. 115, no. 49 (2018): 12359–12364.

8 Liang, H., and Mahadevan, L., 'Growth, Geometry, and Mechanics of a Blooming Lily', *Proceedings of the National Academies of Sciences*, vol. 108, no. 4 (2011): 5516–5521.

9 Noblin, X., Rojas, N.O., Westbrook, J. et al., 'The Fern Sporangium: A Unique Catapult', *Science*, vol. 335 (2012): 1322.

10 Cummins, C., Seale, M., Macente, A. et al., 'A Separated Vortex Ring Underlies the Flight of the Dandelion', *Nature*, vol. 562 (2018): 414–418.

11 Ledda, P.G., Siconolfi, L., Viola, F. et al., 'Flow Dynamics of a Dandelion Pappus: A Linear Stability Approach', *Physical Review Fluids*, vol. 4 (2019): 071901(R).

12 Kim, S., Wu, Z., Esmaili, E. et al., 'How a Raindrop Gets Shattered on Biological Surfaces', *Proceedings of the National Academies of* Sciences, vol. 117, no. 25 (2020): 13901–13907.

13 Ellington, C.P., van den Berg, C., Willmott, A.P. et al., 'Leading-Edge Vortices in Insect Flight', *Nature*, vol. 384 (1996): 626–630.

14 Nabawy, M.R.A., and Crowther, W.J., 'The Role of the Leading-Edge Vortex in Lift Augmentation of Steadily Revolving Wings: A change in Perspective', *Journal of the Royal Society Interface*, vol. 14 (2017): 20170159.

15 Aoyanagi, Y., and Okumura, K., 'Simple Model for the Mechanics of Spider Webs', *Physical Review Letters*, vol. 104 (2010): 038102.

16 Mortimer, B., Soler, A., Wilkins L. et al., 'Decoding the Locational Information in the Orb Web Vibrations of *Araneus Diadematus* and *Zygiella X-notata*', *Journal of the Royal Society Interface*, vol. 16 (2019): 20190201.

17 Su, I., Narayanan, N., Logrono, M.A. et al., 'In Situ Three-dimensional Spider Web Construction and Mechanics', *Proceedings of the National Academies of Sciences*, vol. 118, no. 33 (2021): e2101296118.

18 Dunstan, D.J., and Hodgson, D.J., 'Snails Home', *Physica Scripta*, vol. 89 (2014): 068002.

Chapter 6

1 Retrieved from www.gov.uk/government/statistics/transport-statistics-great-britain-2022/transport-statistics-great-britain-2022-domestic-travel.

2 Ledesma-Alonso, R., and Becerra-Nuñez, 'Electric or Gasoline: A Simple Model to Decide When Buying a New Vehicle', *Environmental Research Communications*, vol. 6 (2024): 025015.

3 Roberson, L.A., Pantha, S., and Helveston, J.P., 'Battery-Powered Bargains? Assessing Electric Vehicle Resale Value in the United States', *Environmental Research Letters*, vol. 19 (2024): 054053.

4 Kerner, B.S., and Rehborn, H., 'Experimental Properties of Phase Transitions in Traffic Flow', *Physical Review Letters*, vol. 79 (1997): 4030.

5 Flynn, M.R., Kasimov, A.R., Nave, J.-C. et al., 'Self-Sustained Nonlinear Waves in Traffic Flow', *Physical Review E*, vol. 79 (2009): 056113.

6 Sugiyama, Y., Fukui, M., Kikuchi, M. et al., 'Traffic Jams Without Bottlenecks: Experimental Evidence for the Physical Mechanism of the Formation of a Jam', *New Journal of Physics*, vol. 10 (2008): 033001.

7 Stern, E.R., Cui, S., Delle Monache, M.L. et al., 'Dissipation of Stop-and-Go Waves via Control of Autonomous Vehicles: Field Experiments', *Transportation Research Part C: Emerging Technologies*, vol. 89 (2018): 205–221.

8 Krapivsky, P.L., and Redner, S., 'Simple Parking Strategies', *Journal of Statistical Mechanics: Theory and Experiment* (2019): 093404.

9   Krapivsky, P.L., and Redner, S., 'Where Should you Park Your Car? The ½ Rule', *Journal of Statistical Mechanics: Theory and Experiment* (2020): 073404.

10 Gershenson, C., and Pineda, L.A., 'Why Does Public Transport Not Arrive on Time? The Pervasiveness of Equal Headway Instability', *PLoS ONE*, vol. 4, no. 10 (2009): e7292.

11 Gershenson, C., 'Self-Organization Leads to Supraoptimal Performance in Public Transportation Systems', *PLoS ONE*, vol. 6, no. 6 (2011): e21469.

12 Carreón, G., Gershenson, C., and Pineda, L.A., 'Improving Public Transportation Systems with Self-Organization: A Headway-Based Model and Regulation of Passenger Alighting and Boarding', *PLoS ONE*, vol. 12, no. 12 (2017): e0190100.

13 Zhang, Y., and Mi, Z., 'Environmental Benefits of Bike Sharing: A Big Data-Based Analysis', *Applied Energy*, vol. 220 (2018): 296–301.

14 Yan, S., Liu, M., and O'Connor, N.E., 'Parking Behaviour Analysis of Shared E-Bike Users Based on a Real-World Dataset: A Case Study in Dublin, Ireland', *2022 IEEE 95th Vehicular Technology Conference* (2022) doi: 10.1109/VTC2022-Spring54318.2022.9860871.

15 Tsushima, H., and Ikeguchi, T., 'Statistical Analysis of Usage History of Bicycle Sharing Systems', *Nonlinear Theory and its Applications*, vol. 13 (2022): 355–360.

16 Gleditsch, M.D., Hagen, K., Andersson, H. et al., 'A Column Generation Heuristic for the Dynamic Bicycle Rebalancing Problem', *European Journal of Operational Research*, vol. 317 (2022): 762–775.

17 Steffen, J., 'Optimal Boarding Method for Airline Passengers', *Journal of Air Transport Management*, vol. 14, no. 3 (2008): 146–150.

18 Steffen, J., 'A Statistical Mechanics Model for Free-for-all Airplane Passenger Boarding', *American Journal of Physics*, vol. 76 (2008): 1114–1119.

19 Steffen, J.H., and Hotchkiss, J., 'Experimental Test of Airplane Boarding Methods', *Journal of Air Transport Management*, vol. 18, no. 1 (2012): 64–67.

20 Retrieved from www.nbcwashington.com/news/national-international/united-airlines-to-make-big-change-to-passenger-boarding-process/3447228.

Chapter 7

1   Retrieved from publications.fifa.com/en/annual-report-2022/finances/2019-2022-cycle-in-review/2019-2022-revenue.

2   Retrieved from www2.deloitte.com/uk/en/pages/sports-business-group/articles/deloitte-football-money-league.html.

3   Farkas, I., Helbing, D., and Vicsek, T., 'Mexican Waves in an Excitable Medium', *Nature*, vol. 419 (2002): 131–132.

4   Darbois Texier, B., Cohen, C., Dupeux, G. et al., 'On the Size of Sports Fields', *New Journal of Physics*, vol. 16 (2014): 033039.

5   Newton, I., 'A Letter of Mr. Isaac Newton, Professor of the Mathematicks in the University of Cambridge; Containing his New Theory About Light and Colors: Sent by the Author to the Publisher from Cambridge, Febr. 6. 1671/72; in order to be

Communicated to the R. Society', *Philosophical Transactions of the Royal Society London* (1672): 3075–3087.

6 Dupeux, G., Le Goff, A., Quéré, D. et al., 'The Spinning Ball Spiral', *New Journal of Physics*, vol. 12 (2010): 093004.

7 Strutt, J.W., 'On the Irregular Flight of a Tennis Ball', *Messenger of Mathematics*, vol. 7 (1877): 14–16.

8 Darbois Texier, B., Cohen, C., Quéré, D. et al., 'Physics of Knuckleballs', *New Journal of Physics*, vol. 18 (2016): 073027.

9 Nelson, N.J., and Strauss, E., 'Dynamical Chaos in a Simple Model of a Knuckleball', *Applied Mathematics and Computation*, vol. 391 (2021): 125661.

Chapter 8

1 Retrieved from www.who.int/publications/m/item/how-to-handwash.

2 Jimenez, J.L., Marr, L.C., Randall, K. et al., 'What Were the Historical Reasons for the Resistance to Recognizing Airborne Transmission During the COVID-19 Pandemic?', *Indoor Air*, vol. 32, no. 8 (2022): e13070.

3 Riley, R.L., Mills, C.C., O'Grady et al., 'Infectiousness of Air from a Tuberculosis Ward. Ultraviolet Irradiation of Infected Air: Comparative Infectiousness of Different Patients', *The American Review of Respiratory Disease*, vol. 85 (1962): 511–525.

4 Riley, R.L., 'What Nobody Needs to Know About Airborne Infection', *American Journal of Respiratory and Critical Care Medicine*, vol. 163 (2001): 7–8.

5 Han, Z.Y., Weng, W.G., and Huang, Q.Y., 'Characterizations of Particle Size Distribution of the Droplets Exhaled by Sneeze', *Journal of the Royal Society Interface*, vol. 10 (2013): 20130560.

6 Pöhlker, M.L., Pöhlker, C., Krüger, O.O. et al., 'Respiratory Aerosols and Droplets in the Transmission of Infectious Diseases', *Reviews of Modern Physics*, vol. 95 (2023): 045001.

7 Mittal, R., Ni, R., and Seo, J-H., 'The Flow Physics of COVID-19', *Journal of Fluids Mechanics*, vol. 894 (2020): F2.

8 Bourouiba, L., 'A Sneeze', *The New England Journal of Medicine*, vol. 375 (2016): e15.

9 Bourouiba, L., Dhandschoewercker, E., and Bush, J.W.M., 'Violent Expiratory Events: On Coughing and Sneezing', *Journal of Fluid Mechanics*, vol. 745 (2014): 537–563.

10 Asadi, S., Wexler, A.S., Cappa, C.D. et al., 'Aerosol Emission and Superemission During Human Speech Increase with Voice Loudness', *Scientific Reports*, vol. 9 (2019): 2348.

11 Abkarian, M., Mendez, S., Xue, N. et al., 'Speech can Produce Jet-like Transport Relevant to Asymptomatic Spreading of Virus', *Proceedings of the National Academy of Science*, vol. 117 (2020): 25237–25245.

12 Stadnytskyi, V., Bax, C.E., Bax, A. et al., 'The Airborne Lifetime of Small Speech Droplets and their Potential Importance in SARS-CoV-2 Transmission', *Proceedings of the National Academy of Science*, vol. 117 (2020): 11875–11877.

13 Bourrianne, P., Xue, N., Nunes, J. et al., 'Quantifying the Effect of a Mask on Expiratory Flows', *Physical Review Fluids*, vol. 6 (2021): 110511.

14 Hunter, P.R., and Brainard, J., 'Changing Risk Factors for Developing SARS-CoV-2 from Delta to Omicron', *PLoS ONE*, vol. 19 (2024) e0299714.

15 Rios de Anda, I., Wilkins, J.W., Robinson, J.F. et al., 'Modeling the Filtration Efficiency of a Woven Fabric: The Role of Multiple Lengthscales', *Physics of Fluids*, vol. 34 (2022): 033301.

16 Sear, R.P., 'Estimating the Population Effects of Non-Pharmaceutical Interventions When Transmission Rates of COVID-19 Vary By Orders of Magnitude From One Contact to Another', *Physical Review E*, vol. 110 (2024): 064302.

17 Li, Y-Y., Wang, J-X., and Chen, X., 'Can a Toilet Promote Virus Transmission? From a Fluid Dynamics Perspective', *Physics of Fluids*, vol. 32 (2020): 065107.

18 Crimaldi, J.P., True, A.C., Linden, K.G. et al., 'Commercial Toilets Emit Energetic and Rapidly Spreading Aerosol Plumes', *Scientific Reports*, vol. 12 (2022): 20493.

19 Goforth, M.P., Boone, S.A., Clark, J. et al., 'Impacts of Lid Closure During Toilet Flushing and of Toilet Bowl Cleaning on Viral Contamination of Surfaces in United States Restrooms', *American Journal of Infection Control*, vol. 52 (2024): 141–146.

Chapter 9

1 Dunbar, R.I.M., 'The Social Brain Hypothesis', *Evolutionary Anthropology*, vol. 6 (1998): 178–190.

2 Sutcliffe, A.J., Dunbar, R.I.M., Binder, J. et al., 'Relationships and the Social Brain: Integrating Psychological and Evolutionary Perspectives', *British Journal of Psychology*, vol. 103 (2012): 149.

3 Dávid-Barrett, T., Kertesz, J., Rotkirch, A. et al., 'Communication with Family and Friends Across the Life Course', *PLoS One*, vol. 11 (2016): e0165687.

4 Dávid-Barrett, T., Diaz, S., Rodriguez-Sickert, C. et al., 'In a Society of Strangers, Kin is Still Key: Identified Family Relations in Large-Scale Mobile Phone Data', arXiv: 2307.03547.

5 Escribano, D., Lapuente, F.J., Cuesta, J.A. et al., 'Stability of the Personal Relationship Networks in a Longitudinal Study of Middle School Students', *Scientific Reports*, vol. 13 (2023): 14575.

6 Saramäki, J., Leicht, E.A., López, E. et al., 'Persistence of Social Signatures in Human Communication', *Proceedings of the National Academy of Sciences*, vol. 111, no. 3 (2014): 942–947.

7 Roy, C., Bhattacharya, K., Dunbar, R.I.M. et al., 'Turnover in Close Friendships', *Scientific Reports*, vol. 12 (2022): 11018.

8 Retrieved from www.pewresearch.org/short-reads/2023/02/02/key-findings-about-online-dating-in-the-u-s.

9 Bruch, E.E., and Newman, M.E.J., 'Aspirational Pursuit of Mates in Online Dating Markets', *Science Advances*, vol. 4 (2018): eaap9815.

10 Tyson, G., Perta, V.C., Haddadi, H. et al., 'A First Look at User Activity on Tinder', arXiv: 1607.01952.

11 Solnyshkov, D., and Malpuech, G., 'Love Might be a Second-Order Phase Transition', *Physics Letters A*, vol. 445 (2022): 128245.

12 Retrieved from www.nimblefins.co.uk/divorce-statistics-uk.
13 Gomberoff, A., Munoz, V., and Romagnoli, P.R., 'The Physics of Custody', *The European Physical Journal B*, vol. 87 (2014): 37.

Chapter 10

1 Liger-Belair, G., Cordier, D., and Georges, R., 'Under-Expanded Supersonic CO_2 Freezing Jets During Champagne Cork Popping', *Science Advances*, vol. 5, no. 9 (2019): eaav5528.
2 Benidar, A., Georges, R., Kulkarni, V. et al., 'Computational Fluid Dynamic Simulation of the Supersonic CO_2 Flow During Champagne Cork Popping', *Physics of Fluids*, vol. 34 (2022): 066119.
3 Wagner, L., Braun, S., and Scheichl, B., 'Simulating the Opening of a Champagne Bottle', *Flow*, vol. 3 (2024): E40.
4 Liger-Belair, G., Bourget, M., Cilindre, C. et al., 'Champagne Cork Popping Revisited Through High-Speed Infrared Imaging: The Role of Temperature', *Journal of Food Engineering*, vol. 116 (2013): 78–85.
5 Retrieved from www.aao.org/newsroom/news-releases/detail/ophthalmologists-warn-flying-champagne-corks-cause.
6 Liger-Belair, G., 'How Many Bubbles in Your Glass of Bubbly?', *Journal of Physical Chemistry B*, vol. 118 (2014): 3156–3163.
7 Atasi, O., Ravisnkar, M., Legendre, D. et al., 'Presence of Surfactants Controls the Stability of Bubble Chains in Carbonated Drinks', *Physical Review Fluids*, vol. 8 (2023): 053601.
8 Poujol, M., Wunenburger, R., Ollivier, F. et al., 'Sound of Effervescence', *Physical Review Fluids*, vol. 6 (2021): 013604.
9 Liger-Belair, G., Khenniche, C., Poteau, C. et al., 'Losses of Yeast-Fermented Carbon Dioxide During Prolonged Champagne Aging: Yes, the Bottle Size Does Matter!', *ACS Omega*, vol. 8 (2023): 22844–22853.
10 Retrieved from www.statista.com/statistics/270275/worldwide-beer-production.
11 Liger-Belair, G., and Cilindre, C., 'How Many CO_2 Bubbles in a Glass of Beer?', *ACS Omega*, vol. 6 (2021): 9672–9679.
12 Lyu, W., Bauer, T., Gatternig, B. et al., 'Experimental and Numerical Investigation of Beer Foam', *Physics of Fluids*, vol. 35 (2023): 023318.
13 Sauret, A., Boulogne, F., Cappello, J. et al., 'Damping of Liquid Sloshing by Foams', *Physics of Fluids*, vol. 27 (2015): 022103.
14 Rodríguez-Rodríguez, J., Casado-Chacón, A., and Fuster, D., 'Physics of Beer Tapping', *Physical Review Letters*, vol. 113 (2014): 214501.
15 Ostmeyer, J., 'Physics of Beer Tapping: Lower vs. Upper Bottle', arXiv: 2002.02896.
16 Sopina, E., Antonescu, I.E., Hansen, T. et al., 'To Beer or Not to Beer: Does Tapping Beer Cans Prevent Beer Loss? A Randomised Controlled Trial', arXiv: 1912.01999.
17 Pereira, L., Wadsworth, F.B., Vasseur, J. et al., 'The Physics of Dancing Peanuts in Beer', *Royal Society Open Science*, vol. 10 (2023): 230376.

Chapter 11

1 Casimir, H.B.G., *Haphazard Reality: Half a Century of* Science (New York: Harper and Row, 1983).

2 Welchman, A.E., Stanley, J., Schomers, M.R., 'The Quick and the Dead: When Reaction Beats Intention', *Proceedings of the Royal Society B*, vol. 277 (2010): 1667–1674.

3 Weller, L., Kunde, W., and Pfister, R., 'Disarming the Gunslinger Effect: Reaction Beats Intention for Co-operative Actions', *Psychonomic Bulletin & Review*, vol. 25 (2018): 761–766.

4 Wang, Z., Xu, B., and Zhou, H-J., 'Social Cycling and Conditional Responses in the Rock-Paper-Scissors Game', *Scientific Reports*, vol. 4 (2014): 5830.

5 Diaconis, P., Holmes, S., and Montgomery, R., 'Dynamical Bias in the Coin Toss', *SIAM Review*, vol. 49, no. 2 (2007): 211–235.

6 Bartoš, F., Sarafoglou, A., Godmann, H.R. et al., 'Fair Coins Tend to Land on the Same Side They Started: Evidence from 350,757 Flips', arXiv: 2310.04153.

7 Bonsma-Fisher, M., and Bonsma-Fisher, K., 'How Big a Table Do You Need for Your Jigsaw Puzzle?', arXiv: 2312.04588.

8 De Ruier, J., Østergaard, E.V., Marker, S. et al., 'Fluid Physics of Telescoping Cardboard Boxes', *Physical Review Fluids*, vol. 7 (2022): 044101.

9 Blasius, B., and Tönjes, R., 'Zipf's Law in the Popularity Distribution of Chess Openings', *Physical Review Letters*, vol. 103 (2009): 218701.

10 Chowdhary, S., Iacopini, I., and Battiston, F., 'Quantifying Human Performance in Chess', *Scientific Reports*, vol. 13 (2023): 2113.

11 Bowling, M., Burch, N., Johanson, M. et al., 'Heads-Up Limit Hold 'em Poker is Solved', *Science*, vol. 347 (2015): 145–149.

12 Moravčík, M., Schmid, M., Burch, N. et al., 'DeepStack: Expert-Level Artificial Intelligence in Heads-Up No-Limit Poker', *Science*, vol. 356 (2017): 508–513.

13 Brown, N., and Sandholm, T., 'Superhuman AI for Multiplayer Poker', *Science*, vol. 365 (2019): 885–890.

14 Kasparov, G., 'Chess, A *Drosophila* of Reasoning', *Science*, vol. 362 (2018): 1087.

15 Silver, D., Hubert, T., Schrittwieser, J. et al., 'A General Reinforcement Learning Algorithm that Masters Chess, Shogi, and Go Through Self-Play', *Science*, vol. 362 (2018): 1140–1144.

16 Schmid, M., Moravčík, M., Burch, N. et al., 'Student of Games: A Unified Learning Algorithm for Both Perfect and Imperfect Information Games', *Science Advances*, vol. 9 (2023): eadg3256.

17 Bakhtin, A., Brown, N., Dinan, E. et al., 'Human-Level Play in the Game of Diplomacy by Combining Language Models with Strategic Reasoning', *Science*, vol. 378 (2022): 1067–1074.

Chapter 12

1 Retrieved from www.statista.com/topics/4679/food-delivery-and-takeaway-market-in-the-united-kingdom-uk.

2 Retrieved from www.businessofapps.com/data/food-delivery-app-market.

3 Alvarado, N.S., and Stevenson, T.J., 'Appetitive Information Seeking Behaviour Reveals Robust Daily Rhythmicity for Internet-Based Food-Related Keyword Searches', *Royal Society Open Science*, vol. 5 (2018): 172080.

4 Retrieved from www.theguardian.com/world/2023/jun/27/pompeii-fresco-find-possibly-depicts-2000-year-old-form-of-pizza.

5 Varlamov, A., Glatz, A., and Grasso, S., 'The Physics of Baking Good Pizza', *Physics Education*, vol. 53, no. 6 (2018): 065011.

6 Haddley, J., and Worsley, S., 'Infinite Families of Monohedral Disk Tilings', arXiv: 1512.03794.

7 Retrieved from roma.repubblica.it/cronaca/2022/09/05/news/gli_spaghetti_a_fuoco_spento_escono_gommosi_lo_chef_stellato_colonna_boccia_in_cucina_il_nobel_parisi-364223344.

8 Retrieved from www.thetimes.co.uk/article/nobel-scientist-urges-italians-save-gas-pasta-cooking-9f3vnsw5p.

9 Retrieved from theconversation.com/italys-pasta-row-a-scientist-on-how-to-cook-spaghetti-properly-and-save-money-191973.

10 Goldberg, N.N., and O'Reilly, O.M., 'Mechanics Based Model for the Cooking-Induced Deformation of Spaghetti', *Physical Review E*, vol. 101 (2020): 013001.

11 Hwang, J., Ha, J., Siu, R. et al., 'Swelling, Softening, and Elastocapillary Adhesion of Cooked Pasta', *Physics of Fluids*, vol. 34 (2022): 042105.

12 Audoly, B., and Neukirch, S., 'Fragmentation of Rods by Cascading Cracks: Why Spaghetti Does Not Break in Half', *Physical Review Letters*, vol. 95 (2005): 095505.

13 Heisser, R.H., Patil, V.P., Stoop, N. et al., 'Controlling Fracture Cascades Through Twisting and Quenching', *Proceedings of the National Academy of Sciences*, vol. 115 (2018): 8665–8670.

14 Retrieved from www.moneybeach.co.uk/the-boardwalk/the-worlds-favourite-takeouts.

15 Kiyama, A., Rabbi, R., Pan, Z. et al., 'Morphology of Bubble Dynamics and Sound in Heated Oil', *Physics of Fluids*, vol. 34 (2022): 062107.

16 Ko, H., and Hu, D., 'The Physics of Tossing Fried Rice', *Journal of the Royal Society Interface*, vol. 17 (2020): 20190622.

17 Liu, L-W., Wang, A-H., Hwang, S-L. et al., 'Prevalence and Risk Factors of Subjective Musculoskeletal Symptoms Among Cooks in Taiwan', *Journal of the Chinese Institute of Industrial Engineers*, vol. 28, no. 5 (2011): 327–335.

18 Pérez-Mohedano, R., Letzelter, N., Amador, C. et al., 'Positron Emission Particle Tracking (PEPT) for the Analysis of Water Motion in a Domestic Dishwasher', *Chemical Engineering Journal*, vol. 259 (2015): 724–736.

Chapter 13

1 Reagan, A.J., Mitchell, L., Kiley, D. et al., 'The Emotional Arcs of Stories are Dominated by Six Basic Shapes', *European Physical Journal: Data Science*, vol. 5 (2016): 31.

2 Del Vecchio, M., Kharlamov, A., Parry, G. et al., 'Improving Productivity in Hollywood with Data Science: Using Emotional Arcs of Movies to Drive Product and

Service Innovation in Entertainment Industries', *Journal of the Operational Research Society*, vol. 72, no. 5 (2020): 1110–1137.

3 Retrieved from www.statista.com/statistics/317408/highest-grossing-film-franchises-series.

4 Roughan, M., Mitchell, L., and South, T., 'How the Avengers Assemble: Ecological Modelling of Effective Cast Sizes for Movies', *PLoS One*, vol. 15, no. 2 (2020): e0223833.

5 Retrieved from www.statista.com/statistics/326011/movie-production-distribution-industry.

6 Lash, M.T., and Zhao, K., 'Early Predictions of Movie Success: The Who, What, and When of Profitability', *Journal of Management Information Systems*, vol. 33, no. 6 (2016): 874–903.

7 Jones, P., 'Diana in the World of Men: A Character Network Approach to Analysing Gendered Vocal Representation in Wonder Woman', *Feminist Media Studies*, vol. 20, no. 1 (2018) 18–34.

8 Gessey-Jones, T., Connaughton, C., Dunbar, R. et al., 'Narrative Structure of a Song of Ice and Fire Creates a Fictional World with Realistic Measures of Social Complexity', *Proceedings of the National Academy of Sciences*, vol. 117, no. 46 (2020): 28582–285588.

9 Vale, R., 'Bayesian Prediction for *The Winds of Winter*', arXiv: 1409.5830.

10 Blum, P., and Wenskat, M., 'Why Monday Never Wins: An Instance of the Secretary Problem', *The Journal of Undergraduate Mathematics and Its Applications*, 14.2 (2021).

INDEX

Note: *italicised* page references indicate illustrations; the suffix 'n' indicates a note